Standardisation of thermal cycling exposure testing

European Federation of Corrosion Publications
NUMBER 53

Standardisation of thermal cycling exposure testing

Edited by
M. Schütze and M. Malessa

Published for the European Federation of Corrosion
by Woodhead Publishing and Maney Publishing
on behalf of
The Institute of Materials, Minerals & Mining

CRC Press
Boca Raton Boston New York Washington, DC

WOODHEAD PUBLISHING LIMITED
Cambridge England

Woodhead Publishing Limited and Maney Publishing Limited on behalf of
The Institute of Materials, Minerals & Mining

Published by Woodhead Publishing Limited, Abington Hall, Abington
Cambridge CB21 6AH, England
www.woodheadpublishing.com

Published in North America by CRC Press LLC, 6000 Broken Sound Parkway, NW,
Suite 300, Boca Raton, FL 33487, USA

First published 2007 by Woodhead Publishing Limited and CRC Press LLC

British Library Cataloguing in Publication Data
A catalogue record for this book is available from the British Library.

Library of Congress Cataloging in Publication Data
A catalog record for this book is available from the Library of Congress.

Woodhead Publishing ISBN 978-1-84569-273-5 (book)
Woodhead Publishing ISBN 978-1-84569-347-3 (e-book)
CRC Press ISBN 978-1-4200-6109-3
CRC Press order number: WP6109
ISSN 1354-5116

The publishers' policy is to use permanent paper from mills that operate a sustainable forestry policy, and which has been manufactured from pulp which is processed using acid-free and elementary chlorine-free practices. Furthermore, the publishers ensure that the text paper and cover board used have met acceptable environmental accreditation standards.

Typeset by Replika Press Pvt Ltd, India
Printed by TJ International Limited, Padstow, Cornwall, England

Contents

Contributor contact details

(* = main contact)

Editors

M. Schütze and M. Malessa
Karl-Winnacker-Institut der
DECHEMA e.V.
D-60061 Frankfurt am Main
Germany

E-mail: schuetze@dechema.de
malessa@dechema.de

Chapter 1

S. Osgerby
National Physical Laboratory
Teddington
TW11 0LW
UK

E-mail: steve.osgerby@npl.co.uk

Chapter 2

R. Pettersson
Swedish Institute for Metals Research
Drottning Kristinas väg 48
S-114 28 Stockholm
Sweden

E-mail: rachel.pettersson@simr.se

Chapter 3

S. Coleman* and D. McGeeney
ISRU
Stephenson Centre
Stephenson Building
University of Newcastle upon Tyne
Newcastle upon Tyne
NE1 7RU
UK

E-mail:
Shirley.Coleman@newcastle.ac.uk
D.V.McGeeney@newcastle.ac.uk

R. Pettersson
Swedish Institute for Metals Research
Drottning Kristinas väg 48
S-114 28 Stockholm
Sweden

E-mail: rachel.pettersson@simr.se

Chapter 4

M. Schütze* and M. Malessa
DECHEMA Gesellschaft für
Chemische Technik und
Biotechnologie e.V.
Theodor-Heuss-Allee 25
60486 Frankfurt am Main
Germany

E-mail: schuetze@dechema.de
malessa@dechema.de

S. Coleman
ISRU
Stephenson Centre
Stephenson Building
University of Newcastle upon Tyne
Newcastle upon Tyne
NE1 7RU
UK

E-mail:
Shirley.Coleman@newcastle.ac.uk

Chapter 5

G. Strehl* and G. Borchardt
Schmidt+Clemens GmbH & Co. KG
ZBTE
Postfach 1140
51779 Lindlar
Germany

E-mail: G.strehl@schmidt-clemens.de

Chapter 6

L. Niewolak* and W. J. Quadakkers
Forschungszentrum Jülich
IWV 2 52425 Jülich
Germany

E-mail: j.quadakkers@fz-juelich.de
l.niewolak@fz-juelich.de

Chapter 7

M. Schütze* and M. Malessa
Karl-Winnacker-Institut der
DECHEMA e.V.
D-60061 Frankfurt am Main
Germany

E-mail: schuetze@dechema.de
malessa@dechema.de

Chapter 8

J. R. Nicholls and T. Rose
School of Industrial and
Manufacturing Science
Cranfield University
Cranfield
MK43 0AL
UK

E-mail: j.r.nicholls@cranfield.ac.uk

Chapter 9

A. Kliewe
MTU Aero Engines GmbH
Dachauer Str. 665
80995 München
Germany

E-mail: Anja.Kliewe@muc.mtu.de

S. Osgerby*
National Physical Laboratory
Teddington
TW11 0LW
UK

E-mail: steve.osgerby@npl.co.uk

Chapter 10

C. Rinaldi*
SSG
CESI RICERCA
via Rubattino, 54
20100 Milano
Italy

E-mail: rinaldi@cesi.it
claudia.rinaldi@cesiricerca.it

H. P. Bossmann

E-mail: hans-
peter.bossmann@power.alstom.com

Chapter 11

M. Mäkipää
BI5 BIOLOGINKUJA 5
PO Box 1000
FIN-02044 VTT
Finland

E-mail: Martti.Makipaa@vtt.fi

Chapter 12

J. R. Nicholls*
School of Industrial and Manufacturing Science
Cranfield University
Cranfield
MK43 0AL
UK

E-mail: j.r.nicholls@cranfield.ac.uk

S. Coleman
ISRU
Stephenson Centre
Stephenson Building
University of Newcastle upon Tyne
Newcastle upon Tyne
NE1 7RU
UK

E-mail:
Shirley.Coleman@newcastle.ac.uk

M. Schütze and M. Malessa
Karl-Winnacker-Institut der DECHEMA e.V
D-60061 Frankfurt am Main
Germany

E-mail: schuetze@dechema.de
malessa@dechema.de

Chapter 13

M. Schütze* and M. Malessa
Karl-Winnacker-Institut der DECHEMA e.V
D-60061 Frankfurt am Main
Germany

E-mail: schuetze@dechema.de
malessa@dechema.de

Appendix

M. Schütze
Karl-Winnacker-Institut der DECHEMA e.V
D-60061 Frankfurt am Main
Germany

E-mail: schuetze@dechema.de

European Federation of Corrosion (EFC) publications: Series introduction

The EFC, incorporated in Belgium, was founded in 1955 with the purpose of promoting European cooperation in the fields of research into corrosion and corrosion prevention. Membership of the EFC is based upon participation by corrosion societies and committees in technical Working Parties. Member societies appoint delegates to Working Parties, whose membership is expanded by personal corresponding membership.

The activities of the Working Parties cover corrosion topics associated with inhibition, education, reinforcement in concrete, microbial effects, hot gases and combustion products, environment sensitive fracture, marine environments, refineries, surface science, physicochemical methods of measurement, the nuclear industry, the automotive industry, computer-based information systems, coatings, tribo-corrosion and the oil and gas industry. Working Parties and Task Forces on other topics are established as required.

The Working Parties function in various ways, e.g. by preparing reports, organising symposia, conducting intensive courses and producing instructional material, including films. The activities of Working Parties are coordinated, through a Science and Technology Advisory Committee, by the Scientific Secretary. The administration of the EFC is handled by three Secretariats: DECHEMA e.V. in Germany, the Société de Chimie Industrielle in France, and The Institute of Materials, Minerals and Mining in the United Kingdom. These three Secretariats meet at the Board of Administrators of the EFC. There is an annual General Assembly at which delegates from all member societies meet to determine and approve EFC policy. News of EFC activities, forthcoming conferences, courses, etc., is published in a range of accredited corrosion and certain journals throughout Europe. More detailed descriptions of activities are given in a Newsletter prepared by the Scientific Secretary.

The output of the EFC takes various forms. Papers on particular topics, for example reviews or results of experimental work, may be published in scientific and technical journals in one or more countries in Europe. Conference proceedings are often published by the organisation responsible for the conference.

In 1987 the then Institute of Metals was appointed as the official EFC publisher. Although the arrangement is non-exclusive and other routes for publication are still available, it is expected that the Working Parties of the EFC will use The Institute of Materials, Minerals and Mining for publication of reports, proceedings, etc., wherever possible.

The name of The Institute of Metals was changed to The Institute of Materials on 1 January 1992 and to The Institute of Materials, Minerals and Mining with effect from 26 June 2002. The series is now published by Woodhead Publishing and Maney Publishing on behalf of The Institute of Materials, Minerals and Mining.

P. McIntyre
EFC Series Editor
The Institute of Materials, Minerals and Mining, London, UK

EFC Secretariats are located at:

Dr B. A. Rickinson
European Federation of Corrosion, The Institute of Materials, Minerals and Mining, 1 Carlton House Terrace, London SW1Y 5DB, UK

Dr J. P. Berge
Fédération Européenne de la Corrosion, Société de Chimie Industrielle, 28 rue Saint-Dominique, F-75007 Paris, FRANCE

Professor Dr G. Kreysa
Europäische Föderation Korrosion, DECHEMA e.V., Theodor-Heuss-Allee 25, D-60486 Frankfurt, Germany

Volumes in the EFC series

1 **Corrosion in the nuclear industry**
Prepared by the Working Party on Nuclear Corrosion

2 **Practical corrosion principles**
Prepared by the Working Party on Corrosion Education (out of print)

3 **General guidelines for corrosion testing of materials for marine applications**
Prepared by the Working Party on Marine Corrosion

4 **Guidelines on electrochemical corrosion measurements**
Prepared by the Working Party on Physico-Chemical Methods of Corrosion Testing

5 **Illustrated case histories of marine corrosion**
Prepared by the Working Party on Marine Corrosion

6 **Corrosion education manual**
Prepared by the Working Party on Corrosion Education

7 **Corrosion problems related to nuclear waste disposal**
Prepared by the Working Party on Nuclear Corrosion

8 **Microbial corrosion**
Prepared by the Working Party on Microbial Corrosion

9 **Microbiological degradation of materials – and methods of protection**
Prepared by the Working Party on Microbial Corrosion

10 **Marine corrosion of stainless steels: chlorination and microbial effects**
Prepared by the Working Party on Marine Corrosion

11 **Corrosion inhibitors**
Prepared by the Working Party on Inhibitors (out of print)

12 **Modifications of passive films**
Prepared by the Working Party on Surface Science and Mechanisms of Corrosion and Protection

13 **Predicting CO_2 corrosion in the oil and gas industry**
Prepared by the Working Party on Corrosion in Oil and Gas Production (out of print)

14 **Guidelines for methods of testing and research in high temperature corrosion**
Prepared by the Working Party on Corrosion by Hot Gases and Combustion Products

15 **Microbial corrosion (Proc. 3rd Int. EFC workshop)**
Prepared by the Working Party on Microbial Corrosion

16 **Guidelines on materials requirements for carbon and low alloy steels for H_2S-containing environments in oil and gas production**
Prepared by the Working Party on Corrosion in Oil and Gas Production

17 **Corrosion resistant alloys for oil and gas production: guidance on general requirements and test methods for H_2S service**
Prepared by the Working Party on Corrosion in Oil and Gas Production

18 **Stainless steel in concrete: state of the art report**
Prepared by the Working Party on Corrosion of Reinforcement in Concrete

19 **Sea water corrosion of stainless steels – mechanisms and experiences**
Prepared by the Working Parties on Marine Corrosion and Microbial Corrosion

20 **Organic and inorganic coatings for corrosion prevention – research and experiences**
Papers from EUROCORR '96

21 **Corrosion–deformation interactions**
CDI '96 in conjunction with EUROCORR '96

22 **Aspects of microbially induced corrosion**
Papers from EUROCORR '96 and the EFC Working Party on Microbial Corrosion

23 **CO_2 corrosion control in oil and gas production – design considerations**
Prepared by the Working Party on Corrosion in Oil and Gas Production

24 **Electrochemical rehabilitation methods for reinforced concrete structures – a state of the art report**
Prepared by the Working Party on Corrosion of Reinforcement in Concrete

25 **Corrosion of reinforcement in concrete – monitoring, prevention and rehabilitation**
Papers from EUROCORR '97

26 **Advances in corrosion control and materials in oil and gas production**
Papers from EUROCORR '97 and EUROCORR '98

27 **Cyclic oxidation of high temperature materials**
Proceedings of an EFC Workshop, Frankfurt/Main, 1999

28 **Electrochemical approach to selected corrosion and corrosion control**
Papers from 50th ISE Meeting, Pavia, 1999

29 **Microbial corrosion (Proc. 4th Int. EFC Workshop)**
Prepared by the Working Party on Microbial Corrosion

30 **Survey of literature on crevice corrosion (1979–1998): mechanisms, test methods and results, practical experience, protective measures and monitoring**
Prepared by F. P. Ijsseling and the Working Party on Marine Corrosion

31 **Corrosion of reinforcement in concrete: corrosion mechanisms and corrosion protection**
Papers from EUROCORR '99 and the Working Party on Corrosion of Reinforcement in Concrete

32 **Guidelines for the compilation of corrosion cost data and for the calculation of the life cycle cost of corrosion – a working party report**
Prepared by the Working Party on Corrosion in Oil and Gas Production

33 **Marine corrosion of stainless steels: testing, selection, experience, protection and monitoring**
Edited by D. Féron on behalf of Working Party 9 on Marine Corrosion

34 **Lifetime modelling of high temperature corrosion processes**
Proceedings of an EFC Workshop 2001, edited by M. Schütze, W. J. Quadakkers and J. R. Nicholls

35 **Corrosion inhibitors for steel in concrete**
Prepared by B. Elsener with support from a Task Group of Working Party 11 on Corrosion of Reinforcement in Concrete

36 **Prediction of long term corrosion behaviour in nuclear waste systems**
Edited by D. Féron and Digby D. Macdonald on behalf of Working Party 4 on Nuclear Corrosion

37 **Test methods for assessing the susceptibility of prestressing steels to hydrogen induced stress corrosion cracking**
Prepared by B. Isecke on behalf of Working Party 11 on Corrosion of Steel in Concrete

38 **Corrosion of reinforcement in concrete: mechanisms, monitoring, inhibitors and rehabilitation techniques**
Edited by M. Raupach, B. Elsener, R. Polder and J. Mietz on behalf of Working Party 11 on Corrosion of Steel in Concrete

39 **The use of corrosion inhibitors in oil and gas production**
Edited by J. W. Palmer, W. Hedges and J. L. Dawson

40 **Control of corrosion in cooling waters**
Edited by J. D. Harston and F. Ropital

41 **Metal dusting, carburisation and nitridation**
Edited by M. Schütze and H. Grabke

42 **Corrosion in refineries**
Edited by J. Harston

43 **The electrochemistry and characteristics of embeddable reference electrodes for concrete**
Prepared by R. Myrdal on behalf of Working Party 11 on Corrosion of Steel in Concrete

44 **The use of electrochemical scanning tunnelling microscopy (EC–STM) in corrosion analysis: reference material and procedural guidelines**
Prepared by R. Lindström, V. Maurice, L. H. Klein and P. Marcus on behalf of Working Party 6 on Surface Science

45 **Local probe techniques for corrosion research**
Edited by R. Oltra on behalf of Working Party 8 on Physico-Chemical Methods of Corrosion Testing

46 **Amine unit corrosion in refineries**
Prepared by J. D. Harston and F. Ropital on behalf of Working Party 15 on Corrosion in the Refinery Industry

47 **Novel approaches to the improvement of high temperature corrosion resistance**
Edited by M. Schütze and W. Quadakkers on behalf of Working Party 3 on Corrosion in Hot Gases and Combustion Products

48 **Corrosion of metallic heritage artefacts: investigation, conservation and prediction of long term behaviour**
Edited by P. Dillmann, G. Béranger, P. Piccardo and H. Matthiessen on behalf of Working Party 4 on Nuclear Corrosion

49 **Electrochemistry in light water reactors: reference electrodes, measurement, corrosion and tribocorrosion issues**
Edited by R.-W. Bosch, D. Féron and J.-P. Celis on behalf of Working Party 4 on Nuclear Corrosion

50 **Corrosion behaviour and protection of copper and aluminium alloys in seawater**
Edited by D. Féron

51 **Corrosion issues in light water reactors: stress corrosion cracking**
Edited by D. Féron and J.-M. Olive on behalf of Working Party 4 on Nuclear Corrosion

52 **(to come)**

53 **Standardisation of thermal cycling exposure testing**
Edited by M. Schütze and M. Malessa

Foreword

The concept of COTEST

The objective of the project was to create the technical basis that will permit the definition, in a comprehensive way, of all the test-related aspects of the cyclic oxidation test. In order to achieve these objectives the following work packages and tasks were undertaken in two phases:

- *Phase I*: Evaluation of the effect of cycling parameters on corrosion behaviour by combining information from the literature and the project contractors with results from an extensive experimental programme.
- *Phase II*: Development and experimental validation of a draft Code of Practice derived from results of phase I and formulation of final version of the Code of Practice.

Phase I of the project consisted of the following steps:

1. Evaluation of the presently used test procedure and experimental facilities for cyclic oxidation testing.
2. Evaluation of cyclic oxidation data available in the literature and from the participating laboratories.
3. Development of a set of test procedures, each adjusted to a wide range of industrial applications.
4. Materials procurement, and modification of test rigs as required.
5. Testing of a series of reference materials under the test conditions defined.

In phase II the following tasks were accomplished:

6. Development of a draft Code of Practice for three types of cyclic oxidation testing based on the outcome of phase I.
7. Experimental validation of the Code of Practice.
8. Formulation of the final version of the Code of Practice and submission to ISO Technical Committee 156.

Each of these steps corresponded to a specific work package as described below.

Work Package 1: Evaluation of the presently used test procedures and experimental facilities for cyclic oxidation testing

The aim of this work package was to obtain an overview of the test parameters in current and historical test procedures in different labs. This review included published literature as well as the test facilities of the contractors. The parameters addressed included

- dwell time at oxidation temperature;
- dwell temperatures;
- cooling time;
- temperature accuracy;
- heating and cooling rate;
- if used, methods used for forced cooling;
- specimen handling and fixation;
- methods of measuring corrosion rate;
- environmental conditions.

The outcome of the work package was used for the definition of the experimental programme in Work Package 3 which offered the main basis for the final definition of the Code of Practice. Details can be found in Chapter 1 by S. Osgerby.

Work package 2: Evaluation of existing cyclic oxidation data available from the literature and those supplied by the various project contractors

The work package aimed to summarise the existing data for various groups of metallic materials with special emphasis on finding correlations between the measured corrosion behaviour of the different materials and the cycling parameters used. The output of this work package was needed for the development of the set of test parameters in Work Package 3 again. The main aim of this evaluation was to

- establish a large database as the starting point for the subsequent test programme;
- derive cycling conditions which are relevant for a large variety of industrial applications;
- estimate the response of different characteristic groups of metallic materials to variations in cyclic oxidation parameters;
- summarise the available scientific background, which permits the description of the effect of changes in test parameters on materials behaviour.

The results of this work package are summarized in Chapters 2 by R. Pettersson and 3 by S. Coleman, D. McGeeney and R. Pettersson).

Work Package 3: Development of a set of test procedures suitable for standardisation

In industrial applications, metallic materials are subjected to different types of thermocycling. Therefore, it would be counter-productive to develop one single standard for the cyclic oxidation testing of all types of materials in all industrial applications. This work package, therefore, aimed at developing three different sets of test procedures, each of which produces data applicable to a range of service conditions, typical of a large number of industrial applications. The three different test types were selected in such a way that together they cover practically the entire range of service conditions in which high-temperature materials are subjected to thermocycling. As a specific feature of the tests, standard statistical analysis methods were used to define confidence limits for sets of corrosion data. Statistical considerations were used to make the testing procedures as efficient as possible to produce maximum useful output.

Three types of tests were addressed:

1. *Long dwell times*. This type of testing aims to simulate conditions in large-scale industrial facilities encountered in applications such as, for example, in power generation plants, the chemical industry, waste incineration plants and the process industry. In these applications the metallic components are designed for extremely long-term operation, e.g. for typically 100 000 hours. Therefore, the time intervals between various thermocyclic cycles are relatively long and the number of cycles is, related to the long operation time of the components, quite small, typically around 50.
2. *Short dwell times*. This type of thermocycling is typically experienced in applications such as industrial gas turbines, aero engines, heat treatment facilities and furnaces. The intervals between start and shutdown of the facilities are generally much shorter than in applications described in test 1. Also the design life and/or the time until complete overhaul/repair (usually 10 000–30 000 hours) are much shorter and, depending on the specific practical application, the number of cycles is much higher than in the cases described in 1.
3. *Ultra-short dwell times*. This type of testing mainly addresses applications of high-temperature alloys as heating elements in the form of wires or foils. Another typical application in which such short cycles prevail would be catalyst foil carriers, e.g. in cars. In such applications the number of cycles is related to the overall design life (often several

hundred to a few thousand hours), extremely high and the time intervals between heating and cooling can be as low as minutes or even seconds. Such conditions are also commonly encountered in a number of other industrial applications such as, for example, burners and hot gas filters but also in a large variety of domestic applications where metallic heating elements are used, e.g. in cooking plates, toasters, boilers, dryers, fryers. For this test a special new test rig was developed and constructed by Cranfield University, based on existing industrial test equipment which was, however, not regarded as ideal.

The test matrix consisted of five characteristic materials (see next work package), three levels of upper dwell time, two levels of lower dwell time, and a comparison of dry and wet conditions. All in all, 36 possible combinations for each material followed from this test matrix including the selection of trials based on statistical concepts. Chapter 4 by M. Schütze, M. Malessa and S. Coleman provides details about this work package. In addition, Chapter 5 by G. Strehl and G. Borchardt provides information about the definition of the hot dwell time.

Work Package 4: Reference materials in the test programme

The number of primary reference materials tested in the COTEST programme was limited to five. This limited number allowed the establishment of a test programme, concentrating on an extensive investigation of the test parameters, which was necessary to be able to define test standards which are able to produce industrial relevant results. The five primary reference materials were selected such that each one is representative for a specific class of high-temperature materials:

Ferritic 9–12% Cr steels which are commonly used as construction materials in power plants and in the chemical and process industries

Typical application temperatures range from about 500 to 750 °C, and the estimated lifetime of components fabricated out of these materials are in most applications in the range of 100 000 hours. As representative material, the high-strength 9% Cr (+ Mo, W) steel P91 was used.

Ferritic 16–18% Cr stabilised steels which are commonly used in the hottest parts of the automotive exhaust lines such as manifolds or in burners

Their oxidation resistance relies on the formation of a protective chromia layer, on the good scale adherence and on the creep properties owing to the addition of stabilisers (Ti, Nb). Typical service temperatures are 700–950 °C. As testing material in the COTEST programme the 18% Cr containing Ti and Nb stabilised AISI441 grade was used.

Austenitic, chromia-forming (Fe, Ni)- and Ni-based materials which are commonly used as construction materials, e.g. in the chemical and process industry, gas turbines and aero engines

For oxidation resistance the alloys rely on the formation of protective chromia based surface scales; however, mostly additional oxide phases, e.g. of the spinel type, are present in the surface scales. Depending on the exact alloy composition, e.g. depending on creep and oxidation resistance of the specific material, these alloys are commonly used in the temperature range 700–1100 °C. In the COTEST project the 32% Ni, 20% Cr-containing austenitic steel Alloy 800H was used.

Nickel-based, γ'-strengthened super alloys primarily designed for components that have to withstand very high mechanical loads

The main alloying elements in these Ni– or Ni(Co)–base alloys are Cr, Al and Ti. Depending on the specific applications, the alloys contain additional γ'-stabilising elements such as Ta, Nb and/or solid solution strengtheners such as W, Mo and Re. The alloys are mainly used as blade and vane materials in stationary gas turbines and aero engines. Typical service temperatures are 850–1100 °C. As a reference material in COTEST the material CM247 was tested.

FeCrAl-based heating element alloys

FeCrAl-based alloys possess far poorer high-temperature strength than the Fe(Ni)- and Ni(Co)-based alloys discussed above. However, the FeCrAl-based materials are frequently used up to ultra-high service temperatures (1000–1400 °C) in cases where mechanical strength is not an important design issue. The excellent high-temperature capability of these materials is related to protective alumina scales which form on the surfaces of the materials during high-temperature service. Typical examples of applications of FeCrAl alloys are heating elements for industrial and domestic applications, catalyst carriers, burners, filters and heat exchangers. In the COTEST project Kanthal A1 was tested as a reference material.

Work Package 5: Experimental investigation of selected materials under cyclic oxidation conditions and evaluation of oxidation behaviour

The main aim of the extensive test programme in this work package was to assess the effect of the cyclic parameters on corrosion behaviour of the five primary reference materials investigated. This work package represented the most extensive part of the COTEST project and the results obtained in this work package are reported in this book. Here only a few key aspects of the results shall be addressed. The first is the definition of a thermocycle in thermocyclic oxidation testing. This was based on a scientific approach where the beginning of the hot dwell period was defined when the temperature reaches 97% of the final upper dwell temperature. This definition is based on kinetics calculations which show that even at a temperature below the final dwell temperature significant oxidation can take place, and that the total amount of initial mass increase during the heating up period can be represented by assuming that the hot dwell time starts at 97% of the upper dwell temperature (for details see Chapter 5). The end of the hot dwell period is reached as soon as the temperature decreases (usually by removing the furnace from the specimen or vice versa) and the cooling period ends once a temperature of 50 °C has been reached. The latter value has been set somewhat arbitrarily but was based on the assumption that in no laboratory in the world ambient temperatures exceed 50 °C. Another important observation in the tests was that a test duration of at least 300 hours should be used in order to show differences in the spalling or oxidation behaviour of the materials. In the frame of this work package the evaluation procedure of the test results was also developed. This procedure aims at quantifying the oxidation rate constant k_n, the time of protective oxidation behaviour (to the beginning of spalling or breakaway oxidation), $t_{\text{protective}}$, and a mass change value determined after test completion describing the oxidation behaviour after the end of the protective period. It is described in detail in the Code of Practice. As the evaluation shows, the k_n values (i.e. the oxidation behaviour in the protective period) do not depend on the cycling parameters. This is not a surprising result since, in the protective time range, oxidation is solely governed by diffusion of the oxide-forming species in the scales. This situation changes once the protective time range is exceeded and in particular the values for the mass change at the end of the test depend greatly on the cycling parameters as does the length of the protective period $t_{\text{protective}}$.

The significance of the influence of the different test parameters on these oxidation values has been evaluated by statistical methods. It turns out that the cyclic oxidation behaviour described by these parameters depends to a large extent on the type of material used, i.e. whether Fe-Cr-spinels, chromia or alumina scales are formed. This can be regarded as an indication that the test procedure is powerful enough to distinguish between oxide scales of different protective potential.

Details of the experimental results are given in Chapters 6 by L. Niewolak and W. J. Quadakkers, 7 by M. Malessa and M. Schütze, 8 by J. R. Nicholls, 9 by A. Kliewe and S. Osgerby, 10 by C. Rinaldi and H. P. Bossmann and 11 by M. Mäkipää.

Work Package 6: Development of a draft Code of Practice

Based on the outcome of the experimental studies in Work Package 5, in combination with the evaluation from Work Package 2, a draft Code of Practice for thermocyclic oxidation testing was developed. This Code of Practice was formulated along the guidelines used for putting up ISO test standards. Therefore, it addresses the following items:

- scope of the Code of Practice;
- normative references;
- definitions;
- test apparatus (design, temperature monitoring, gas supply);
- test pieces (size and shape, characterisation prior to testing);
- test method (definition of a thermocycle, types and dwell times of thermocycles, test duration, supporting of test pieces, test environment in air oxidation testing, test parameters in complex corrosive environments, determination of mass change by oxidation);
- post-test evaluation of test pieces (macroscopic evaluation, metallographic cross-section);
- reporting.

Work Package 7: Experimental validation of the draft Code of Practice

All principal contractors and a number of subcontractors were involved in the experimental validation of the draft code of practice defined in Work Package 6. The tests were organised in such a way that each test was carried out in several laboratories. After testing, the contractors were requested to supply the outcome of the materials behaviour by:

- exact record of all test parameters during the course of the test;
- gravimetric data for specimens as function of time;
- gravimetric data for spalled oxide as function of time;
- metallographic examination of corrosion damage;
- characterisation of corrosion products by electron microscopy on selected specimens.

Since all tests had to follow the guidelines developed in the COTEST programme it was expected that no large scatter should occur between the results from the different laboratories. It turned out at the end of this work package that indeed in most cases there was only very little scatter between results from different laboratories and that this scatter could be explained mainly by the different cooling rates inherent to the different furnaces. However, even the duration of the protective time range and the behaviour in the non-protective time range in many cases showed a reasonable degree of agreement among the different laboratories. It was therefore concluded at the end of Work Package 7 that the guidelines developed so far are sufficiently powerful for industrial use and will lead to reliable data with a good degree of intercomparison between different laboratories. The results of the validation testing are presented in Chapter 12 by J. R. Nicholls, S. Coleman, M. Malessa and M. Schütze.

Work Package 8: Formulation and fine-tuning of the final version of the Code of Practice

After experimental evaluation of the draft Code of Practice, the code was slightly amended. This was mainly directed towards additional details in the test procedure which were detected as significant in Work Package 7. All in all, however, there was no need for a substantial revision of the draft Code of Practice developed in Work Package 6. The final version of the Code of Practice has already been submitted to Work Group 13 of ISO Technical Committee 156 which started first work in November 2005. The work item was finished by the end of 2006 and is now in the voting process at the International Organization for Standardization (ISO).

Concluding remarks

In the joint European COTEST project a solid basis has been developed for a reliable and meaningful standard for high-temperature cyclic oxidation testing. This basis combines scientific approaches with industrial needs and for the first time allows an intercomparison of data from different laboratories. In the following chapters details about the work within the COTEST project and the results are reported.

Preface

In modern high-temperature technology, materials play a key role with respect to performance, reliability, safety, economic profit and ecological compatibility. The advances in the development of energy conversion systems (low CO_2 emission fossil fuel-fired power stations, solid oxide fuel cells, waste and biomass combustion or gasification, coal conversion, etc.) and in engines for transportation (car engines, catalytic converters, advanced jet engines, etc.) are to a large extent based on reliable long-term performance of high-temperature materials. During operation of such high-temperature technologies these materials are subjected to a complex interaction of temperature changes, oxidative and corrosive high-temperature attack and mechanical stresses. This interaction determines whether components exhibit premature failure or show reliable and safe long-term performance and it also limits the upper service temperature, which decides the degree of efficiency and hence the economical and ecological performance of such a plant. A key role within this interaction is played by high-temperature oxidation and/or corrosion and it is somewhat surprising that although issues of high-temperature corrosion have been dealt with for almost a century in science and in industry, no widely used standards or guidelines exist with regard to reliable testing under such conditions.

Among all the methods for high-temperature corrosion testing, the thermal cycling oxidation test, also known as cyclic oxidation test, has become the most widely used in industry with regard to the number of specimens tested. However, each company and each research institute uses its own modification of this type of test with different test parameters so that in the end no intercomparison of the results from different laboratories is possible. A set of standards or a code of practice which could be used by laboratories does not yet exist. A recent European workshop revealed that in particular industry has a very strong interest in the development of a standard for cyclic oxidation testing in order to get reliable and intercomparable data from such tests for design as well as for new alloy development programmes.[1]

[1] M. Schütze, W. J. Quadakkers (Eds.), *Cyclic Oxidation of High Temperature Materials,* EFC monograph number 27, Institute of Materials, London **1999**.

Owing to the number of parameters influencing materials behaviour under these conditions, the large number of users of this test and the different variants of tests used presently, it would have been impossible for a small group to work on a solution of this problem. The expertise in this field is scattered in industrial and scientific research laboratories all over the world. This was the reason why, following the above-mentioned workshop, a cyclic oxidation testing initiative group was formed whose aim was to work towards the establishment of a respective standard.

It was, however, realised that, because of the situation then, prenormative research was necessary in order to provide a basis for such a set of standards. Being aware of this situation, the European Commission issued a dedicated call 'Measurement and testing – methodologies to support standardisation in community policies' within the Framework V GROWTH programme addressing this specific problem.[2] Following this call the European project 'Cyclic oxidation testing – Development of a code of practice for the characterisation of high temperature materials performance (COTEST)' was started with partners from 11 European countries.

The main technical and scientific objectives of the project, which lasted from January 2002 to December 2004, were:

- to quantify the role of the test parameters that lead to scatter between the results of different laboratories;
- to develop a reliable and meaningful test procedure for the cyclic oxidation test with three variants, to account for the most common technical applications;
- to draft a Code of Practice based on the results of the project for submission to ISO/TC 156 as part of the work item 'Cyclic oxidation testing' of Work Group 13.

The present European Federation of Corrosion (EFC) volume describes in detail the results of the different work packages of the COTEST project. The first part of this volume focuses on the situation prior to COTEST, while in the second part the experimental results of the project are described in detail. The last part covers the validation testing of a draft version of the Code of Practice that was developed in the project. The final version of the Code of Practice is also given in the last part. However, this version of the Code of Practice is currently under the revision of ISO TC156 WG 13.

The authors report work that has been performed within in the frame of the COTEST project and thus provide a comprehensive survey of the influence

[2] http://europa.eu.int/comm/research/growth

of test parameters on thermal cycling oxidation behaviour. The contribution of each author is gratefully acknowledged. Special thanks are also given to the European Commission for financial furtherance of the COTEST project.

M. Schütze
Karl-Winnacker-Institut der DECHEMA e.V., Frankfurt, Germany
Chairman of EFC Working Party 'Corrosion by Hot Gases and Combustion Products'

M. Malessa
Karl-Winnacker-Institut der DECHEMA e.V., Frankfurt, Germany

Acknowledgement

The COTEST project was funded by the European Commission in the *Measurement and Testing Activities* of the Framework V programme under the project no. G6RD-CT-2001-00639, which is gratefully acknowledged by all project participants.

Part I

Methods and procedures in thermal cycling oxidation testing prior to COTEST

1
Survey of existing test procedures and experimental facilities

S. OSGERBY, National Physical Laboratory, UK

1.1 Introduction

Cyclic oxidation testing is a key method to aid material selection and to predict service lifetime of components. However, it is a complex procedure that has many possible variables. Hence it is often difficult to compare data from different laboratories unless either (i) the procedures in each laboratory are similar or (ii) the influence of any differences in procedure on the resultant data is known.

The first task within the COTEST project was thus to survey existing facilities in order to establish current practice. The basic requirements of a cyclic oxidation facility are a zone for heating the specimens, a zone for cooling the specimens (this may be separate from the heating zone but is not necessarily so) and a method to transport the specimens between these two zones. Initial discussions within the COTEST consortium resulted in an agreed list of those parameters that were likely to differ between laboratories/facilities. These were classified into six categories:

1. Type of layout.
2. Temperature control.
3. Heating/cooling practice.
4. Atmosphere.
5. Specimens.
6. Measurement techniques.

At this stage it was also decided to classify cyclic oxidation testing into three classes based upon the time at temperature during each cycle. The definitions were agreed as:

1. Long cycle: $t > 3.5$ hours.
2. Short cycle: 10 min $< t < 3.5$ hours.
3. Ultra-short cycle: $t < 10$ min.

A questionnaire in the form of an MS-Excel spreadsheet was distributed to the partners. This requested details of the test procedure and examples of test data. The form of the questionnaire allowed easy collation of returned data into an appropriate database. Descriptions of 22 test facilities and operating procedures were received. Not all the required information was available on every facility but sufficient information was forthcoming to allow conclusions to be drawn. The data supplied were analysed for differences in the six categories above. Each category was analysed according to whether the predominant use of the facility was for long, short or ultra-short cycle times.

1.2 Summary of layout

The first aspect to consider is the overall layout of the cyclic oxidation facility. The results are summarised in Table 1.1. The numbers of facilities that had a vertical and those that had a horizontal arrangement for the heating and cooling chambers were approximately equal. The 'other' category included a burner rig, a muffle furnace and a direct resistance heating facility for ultra-short cycle testing.

The choice of open or closed system is summarised in Table 1.2. Most facilities are 'closed': an arrangement that prevents any influence of local environment, e.g. chlorine in the atmosphere at laboratories near the coast, affecting the test data.

1.3 Temperature control

In most of the facilities, temperature was controlled and monitored by thermocouple. One piece of equipment, a direct heating system for ultra-

Table 1.1 Summary of overall layout of test facilities

Orientation	Long cycle	Short cycle	Ultra-short cycle	Total
Vertical	3	7	–	10
Horizontal	2	5	–	7
Other	1	1	1	3
Not specified	1	–	–	1

Table 1.2 Summary of open/closed choice for test facility

Open or closed system	Long cycle	Short cycle	Ultra-short cycle	Total
Open	1	4	–	5
Closed	5	8	–	13
Not specified	1	1	1	3

short cycle testing, was controlled by pyrometer. Thermocouples of types R and S were the most commonly used, although types K and B were also used. This information is summarised in Table 1.3.

The positioning of thermocouples is summarised in Table 1.4. One facility attached a thermocouple to a dummy specimen in the assembly: this is the most accurate way of measuring the surface temperature of specimens and it is perhaps surprising that more laboratories do not employ this practice. The majority of the facilities used thermocouples that were placed near to the specimen assembly while three laboratories relied upon the furnace controller. This latter practice is potentially misleading as the furnace tube and specimens will not be in thermal equilibrium for some time after the start of any thermal cycle – a calibration exercise is strongly recommended if this practice is used.

Table 1.5, shows a wide range in the calibration period for thermocouples. This period should be governed by the operating temperature and thermocouple

Table 1.3 Summary of thermocouple type

Thermocouple type	Long cycle	Short cycle	Ultra-short cycle	Total
B	1	–	–	1
K	–	3	–	3
R	1	3	–	4
S	3	5	–	8
Unspecified	2	2	–	4

Table 1.4 Summary of thermocouple positioning

Thermocouple position	Long cycle	Short cycle	Ultra-short cycle	Total
Attached to dummy specimen	–	1	–	1
Adjacent to specimens	4	8	–	12
Furnace controller	1	2	–	3
Unspecified	2	2	–	4

Table 1.5 Summary of thermocouple calibration periods

Calibration period	Long cycle	Short cycle	Ultra-short cycle	Total
Before each test	1	1	–	2
500 h usage	–	1	–	1
6 months	1	1	–	2
1 year	–	6	–	6
2 years	1	–	–	1
Not specified	4	4	–	8

type; therefore it is not surprising that there is a wide spread between laboratories, which may be performing tests under different conditions. Best practice is, of course, to calibrate before each test but experience may be sufficient for laboratories to relax this procedure as data on thermal drift are recorded and analysed. All thermocouples require recalibration at intervals and this issue was identified as one that should be addressed in the Code of Practice to be produced at the end of the COTEST project.

The temperature stability that is required is a function of the upper temperature. The guidelines generated during the TESTCORR [1] project are sufficient to define the requirements of cyclic oxidation testing. The data generated during this survey are summarised in Table 1.6 and in general are adequate for this type of testing.

1.4 Heating/cooling practice

The majority of the facilities move the specimens from the hot to the cool zone of the test facility. Other methods employed include moving the furnace; opening the furnace door; and programming, or simply switching off the furnace controller. The information supplied is summarised in Table 1.7.

The methods used to accelerate cooling are summarised in Table 1.8. The vast majority of facilities used either natural cooling or a gas blast, with the former being slightly more popular. Two facilities were cycled by means of programming the furnace controller. The type of cooling has a direct influence

Table 1.6 Summary of temperature stability during hold period

Temperature stability ± °C	Long cycle	Short cycle	Ultra-short cycle	Total
<1	–	1	–	1
1–2	–	5	–	5
2–3	1	–	–	1
3–5	2	5	–	7
>5	1	1	–	2
Not specified	3	1	–	4

Table 1.7 Summary of cycling methods used

Cycling method	Long cycle	Short cycle	Ultra-short cycle	Total
Specimen moved	3	10	–	13
Furnace moved	–	1	–	1
Furnace opened	–	1	–	1
Furnace programmed	1	–	–	1
Furnace switched off	1	–	–	1
Not specified	2	–	1	3

on the cooling rate that can be achieved, which is the next test parameter to be addressed.

Only ~30% of the responses included information regarding cooling rates and within these data there is a wide range of values reported (Table 1.9). The slowest cooling rates were generated using programmed furnace controllers; however, it should be noted that in these tests the slow cooling rates were intended.

1.5 Atmosphere

The majority of tests are carried out in laboratory air with few facilities controlling humidity. Some laboratories used bottled gas mixtures; these tests were primarily for specific applications or investigating oxidation/corrosion mechanisms. The data are summarised in Table 1.10. The figures in parentheses indicate the number of facilities with controlled humidity.

The questionnaire returns for gas flow rate are summarised in Table 1.11. These rates were generally not specified. This is understandable for tests in laboratory air where the atmosphere is static. As with temperature stability,

Table 1.8 Summary of cooling methods

Cooling method	Long cycle	Short cycle	Ultra-short cycle	Total
Natural	3	5	1	9
Gas blast	1	6	–	7
Other	3	1	–	4
Not specified	1	1	–	2

Table 1.9 Summary of reported instantaneous cooling rates

Cooling rate (K/s)	Long cycle	Short cycle	Ultra-short cycle	Total
0.01–0.1	1	–	–	1
0.1–1	–	1	–	1
1–10	1	1	–	2
>10	–	3	–	3
Not specified	5	7	1	13

Table 1.10 Summary of test atmospheres used

Atmosphere	Long cycle	Short cycle	Ultra-short cycle	Total
Laboratory air	4 (1)	8 (1)	1	13 (2)
Synthetic air	–	4 (1)	–	4 (1)
Other	2	1	–	3
Not specified	–	–	–	–

the recommendations developed in the TESTCORR project [1] should be adopted directly for cyclic oxidation testing.

1.6 Test pieces – geometry, preparation and handling

The majority of laboratories used rectangular test pieces. Cylindrical test pieces were used in short cycle tests, particularly for coatings where edge effects were considered to be important. One laboratory tested small components, as well as specimens in their facility. The data are summarised in Table 1.12.

The size/surface area of test pieces is influential in controlling the cooling rate during thermal cycling. A wide range of specimen sizes and surface areas were used (Table 1.13. This test parameter was identified as needing to be addressed during the COTEST project.

The surface condition of test pieces is recognised as an important test variable. The surface finishes used by the different laboratories are summarised in Table 1.14. Coated test pieces and components are tested in the as-received condition, while uncoated test pieces usually undergo some preparation, the final finish being dependent upon material and application (Table 1.15).

Table 1.11 Summary of gas flow rates

Flow rate (m/s)	Long cycle	Short cycle	Ultra-short cycle	Total
10^{-5}–10^{-4}	–	1	–	1
10^{-4}–10^{-3}	–	1	–	1
10^{-3}–10^{-2}	2	2	–	4
Not specified	5	9	1	15

Table 1.12 Specimen geometry

Geometry	Long cycle	Short cycle	Ultra-short cycle	Total
Rectangular	5	8	–	13
Cylinder	–	4	1	5
Disk	2	–	–	2
Other	–	1	–	1
Not specified	–	–	–	–

Table 1.13 Range of test piece surface areas used in cyclic oxidation testing

Surface area (mm^2)	Long cycle	Short cycle	Ultra-short cycle
Maximum	680	3117	189
Minimum	225	280	189

Table 1.14 Specimen surface finish

Final finish	Long cycle	Short cycle	Ultra-short cycle	Total
100/120 grit	–	3	–	3
180 grit	–	1	–	1
600 grit	1	3	–	4
1200 grit	3	1	–	4
2500 grit	1	–	–	1
1 μm polish	1	1	–	2
As–received	–	4	1	5

Table 1.15 Specimen cleaning procedure

Method	Long cycle	Short cycle	Ultra-short cycle	Total
Ultrasonic	4	5	–	9
Solvent	–	2	1	3
Other	1 (LEAFA*)	–	–	1
Not specified	1	6	–	7

*Life extension of alumina forming alloys.

A wide range of methods to support the test pieces were used, summarised in Table 1.16. The materials used to construct these supports were typically quartz, platinum or glass fibre for hooks, alumina or quartz for crucibles; and metallic alloys (oxide dispersion strengthened (ODS) or FeCrAl) for cages. The method used to support the test pieces affects the thermal mass of the assembly and thus will influence the cooling rate during thermal cycling.

1.7 Measurement/evaluation techniques

A wide range of evaluation techniques were used. Individual laboratories often use more than one in a test programme. The use of the various techniques is summarised in Table 1.17. The most common method was the net mass change determined by intermittent weighing of the test pieces [2]. Some laboratories also collected the spalled scale to determine gross mass change. Post-test evaluation often included determination of the scale composition using scanning electron microscopy-energy dispersive X-ray analysis (SEM-EDX).

1.8 Conclusions

The data supplied from project partners indicate that there is a wide range in experimental practice. Thus the need for a standard for this type of testing is confirmed.

Table 1.16 Summary of test piece support systems

Method	Long cycle	Short cycle	Ultra-short cycle	Total
Hook	3	1	–	4
Crucible	3	6	–	9
Cage	–	3	–	3
Other	–	1	1	2
Not specified	1	1	–	2

Table 1.17 Measurement techniques

Measurement technique	Long cycle	Short cycle	Ultra-short cycle	Total
Mass change (net)	6	8	–	14
Mass change (gross)	3	3	–	6
Mass change (after descaling)	1	1	–	2
Scale thickness	–	1	–	1
Metal loss	–	1	–	1
Scale composition	–	5	–	5
Other	–	3	1	4

1.9 References

1. A B Tomkings, J R Nicholls and D G Robertson, *Discontinuous Corrosion Testing in High Temperature Gaseous Atmospheres*, TESTCORR, Final Report EUR/19479 European Commission 2001.
2. J R Nicholls, Discontinuous Measurements of High Temperature Corrosion, in *Guidelines for Methods of Testing and Research in High Temperature Corrosion*, eds H J Grabke and D B Meadowcroft, Institute of Materials, London 1995 pp 11–36.

2
Compilation of cyclic oxidation data

R. PETTERSSON, SIMR, Sweden

2.1 Introduction

This work was conducted to assess the available cyclic oxidation data at the start of the COTEST project. This included an analysis of the cyclic test conditions employed and an assessment of the effects of test parameters. The required materials data parameters for Work Package 2 were discussed at the initial meeting and a series of questions developed together with the UK National Physical Laboratory (NPL) and Industrial Statistics Research Unit (ISRU). The questionnaire mentioned in Chapter 1 was circulated to all partners and preliminary analysis of all data sets was performed. Any uncertainties in data were verified by direct contact with the partners who submitted the data. Close interaction was maintained throughout with ISRU, which was responsible for statistical analysis.

2.2 Cycle length and test duration

A total of 139 data sets were reported for the structural materials categories covered by COTEST. The cycle length was classified on the basis of the dwell time at the upper temperature into ultra-short, short or long. The boundaries for these categories were redefined during analysis to give better consistency with reported data and envisaged testing within the project. The final boundaries gave clear divisions and good agreement with prior usage (see Fig. 2.1):

- Ultra-short: $t \leq 10$ min (reported data 1 min and 2 min).
- Short: 10 minutes $< t \leq 8$ h (reported data 20 min, 1, 2, $3^1/_2$h).
- Long: $t > 8$ h (reported data 20, 24, 96, 100, 166 and 200 h).

The number of cycles employed is shown in Table 2.1. The median values of the time at temperature were 636 h for long cycle testing, 930 h for short cycle and 500 h for ultra-short cycle. This gives an indication of the test duration which the participants have found necessary in order to assess the

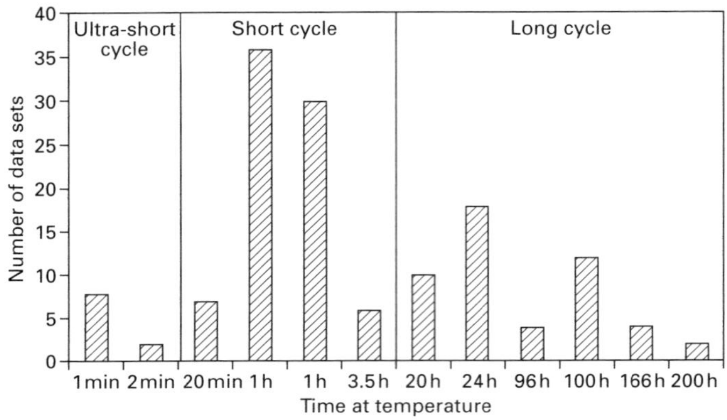

2.1 Cycle duration for all 139 reported data series and division into categories.

Table 2.1 Number of cycles employed for submitted data sets

Number of cycles	Long cycle	Short cycle	Ultra-short cycle	Total
1–9	10	–	–	10
10–99	31	11	–	42
100–499	3	26	–	29
500–999	1	18	–	19
≥1000	5	24	10	39
Total	50	79	10	139

cyclic high-temperature performance of materials, and as such provides a basis for future recommendations.

The techniques employed for evaluation of the cyclic high-temperature oxidation data are shown in Table 2.2. By far the most common method was recording of the net mass change as a function of time by intermittent weighings. Over half of the data sets also involved collection of the spalled scale to permit separate recording of the gross weight change. There were no instances in which weight measurements were used for ultra-short cycle testing; the usual parameter in this case was the number of cycles to failure of thin specimens. Metallographic evaluation of cross-sections was employed in an appreciable number of cases for all three types of testing. Post-test evaluation often also included determination of the scale composition using scanning electron microscopy (SEM).

2.3 Materials and environments

The reported data was fairly well divided between the five categories of structural materials covered by COTEST: austenitic Ni-Cr steels, ferritic steels (9–12% Cr), ferritic steels (16–18% Cr), FeCrAl(RE) and Ni–base alloys (Table 2.3). Because of the relative paucity of data for the ferritic steels, the two original categories, 9–12% Cr and 16–18% Cr, were combined. The largest group, almost half of the submitted data sets, was for nickel–base alloys. Eight data sets for nickel–base alloys were for coated materials and thus strictly outside the scope of this work, but are included in the analysis for purposes of comparison.

More than half of the reported tests were carried out in laboratory air without facilities for controlling humidity (Table 2.4). This means that large variations in water vapour level are to be expected, depending on the geographic location of the site, the time of year and the use of air conditioning/heating in the laboratory. These factors make this type of testing unsuitable for

Table 2.2 Measurement techniques used for evaluation of submitted data sets. A number of sets include several evaluation methods

Measurement technique	Long cycle	Short cycle	Ultra-short cycle	Total
Mass change (net)	36	60	–	96
Mass change (gross)	22	28	–	50
Mass change (after descaling)	–	7	–	7
Metallographic section	15	34	8	59
Rate constant only	2	–	–	2
End point only	8		10	18

Table 2.3 Summary of reported testing data, by cycle length

	Austenitic	Ferritic	FeCrAl(RE)	Ni–base	Coated Ni–base	Total
Ultra-short	–	–	2	8	–	10
Short	15	9	3	46	6	79
Long	8	6	25	9	2	50
Total	23	15	30	63	8	139

Table 2.4 Summary of reported data, by material type and environment

	Austenitic	Ferritic	FeCrAl(RE)	Ni–base	Coated Ni–base	Total
Dry air	13	4	8	–	–	25
Lab air	6	5	12	49	–	72
Moist air	–	–	6	4	8	18
Corrosive	4	6	2	10	–	22
Unspecified	–	–	2	–	–	2
Total	23	15	30	63	8	139

standardisation. Some laboratories used bottled synthetic air, which was used either dry or after controlled humidification. Only two corrosive gas mixtures were reported: Ar–5%HCl and Ar–5%H_2–1%H_2S. Each of these was used for a number of tests in the same two laboratories.

2.4 Variability of results

An understanding of the normal variability between replicate samples forms a vital background against which the effect of experimental variables should be assessed. One multiple data set submitted to the COTEST survey is shown in Fig. 2.2. This indicates a relatively small variation in the oxidation rate, as seen at 850 °C. However, the time to spallation, seen in the data at 950 °C, shows much larger specimen-to-specimen variability.

2.5 Influence of experimental variables

Intercomparison of data from different sources was in most cases not possible because of the simultaneous variation in a number of test parameters. However, data from a single source was amenable to analysis since several closely

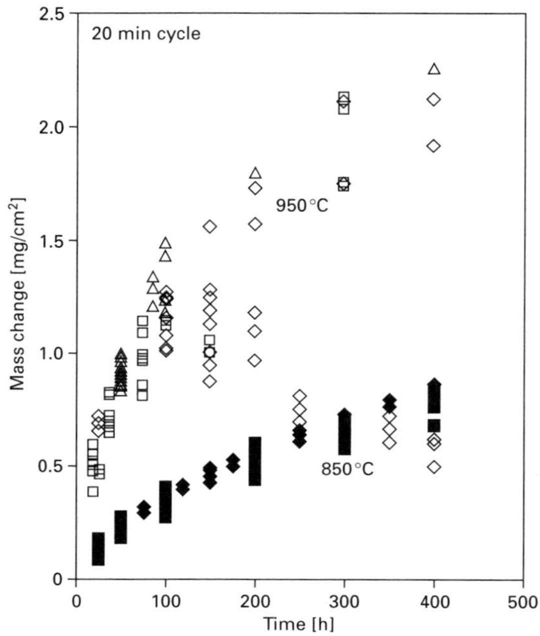

2.2 Results of multiple experiments on a ferritic 441 steel in laboratory air at two temperatures, showing extent of specimen-to-specimen variability (Ugine-Savoie).

allied data sets were frequently reported. The primary variables investigated were alloy composition and temperature, but systematic studies were also reported on the effect of the following variables:

- Specimen thickness for FeCrAl(RE) alloys. This is illustrated in Fig. 2.3 and is due to more rapid exhaustion of the matrix aluminium for thinner specimens.
- The use of natural or enforced cooling. The data for two nickel–base alloys in Fig. 2.4 indicates that spallation is promoted by rapid cooling, which is expected to create larger thermal stresses in the oxide.
- The effect of cycle duration for FeCrAl(RE) alloys (Fig. 2.5). The effect cannot be unequivocally identified on the basis of the submitted data because of large specimen-to-specimen variation.

2.6 Conclusions

The materials data submitted within Work Package 2 supports the relevance of the parameters suggested for systematic evaluation in Work Package 5 of COTEST. At the same time it underlines the necessity of this evaluation because little of the reported data from different sources is amenable to

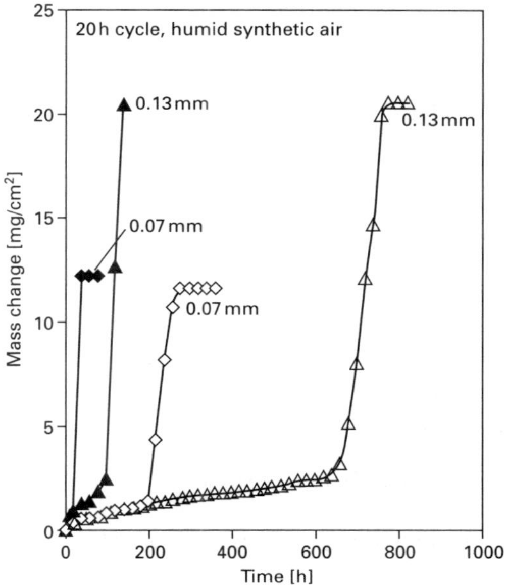

2.3 Effect of specimen thickness on time to breakaway for FeCrAl(RE) alloys at 1300 °C (solid points) and 1200 °C (open points) (University of Liverpool).

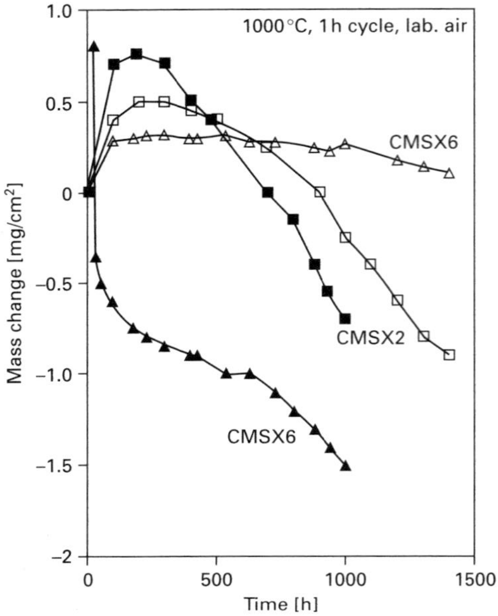

2.4 Effect of natural cooling (open points) or gas blast cooling (solid points) on cyclic oxidation of two nickel–base alloys (KWI-Dechema).

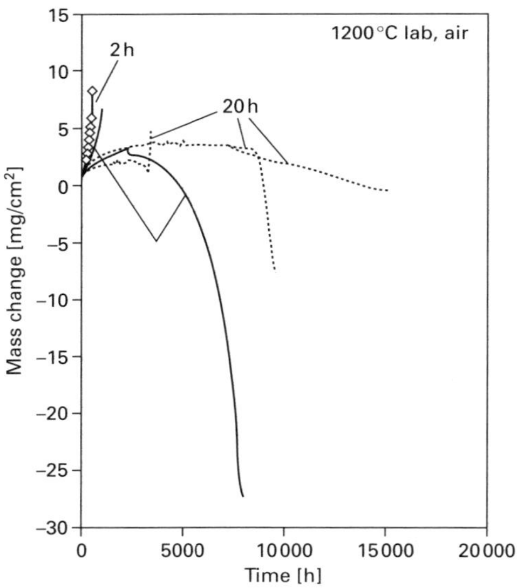

2.5 Effect of cycle duration on cyclic oxidation of PM2000 (FZJ).

intercomparison. The role of testing variables on the outcome of cyclic oxidation testing was thus not sufficiently clear at this stage of the project to permit standardisation of testing procedures.

3

Statistical analysis of cyclic oxidation data

S. COLEMAN and D. MCGEENEY,
Newcastle University, UK and
R. PETTERSSON, SIMR, Sweden

3.1 Test parameters

It has been possible to investigate the effects of some test parameters because there are replicates at differing levels. Others, such as wet/dry air which have an effect on oxidation rate (see, for example, Nicholls and Bennet, 2000), could not be assessed.

3.2 Comparing and summarising the mass change data

The output profiles can be compared visually by looking at Figs 3.1–3.27. There are distinct differences in some of the profiles and, where there is only one test parameter varying, it may be reasonable to attribute the differences to that factor. In some cases two test parameters vary simultaneously and their effects cannot be distinguished. Further analysis can be carried out to confirm significant relationships and also to use the data scatter to estimate the experimental error, which can be used to indicate the number of replications required for future experiments.

The output profiles are summarised by a few key oxidation parameters rather than dealing with the whole profile, which would be difficult to interpret. The various stages of oxidation can be detected from the output profiles if there are sufficient measurements at the appropriate time points. The output profiles here range from single measurements (IDs 61–68 in Fig. 3.9 and 78–85 in Fig. 3.17) to 4096 measurements (ID 94 not plotted). Estimating oxidation parameters, such as parabolic rate constant from the data is open to subjective judgement.

3.3 Statistical analysis – within sources

Only the most important factors show up as significant in statistical analysis when the number of cases is small. Analysis of variance (ANOVA) with

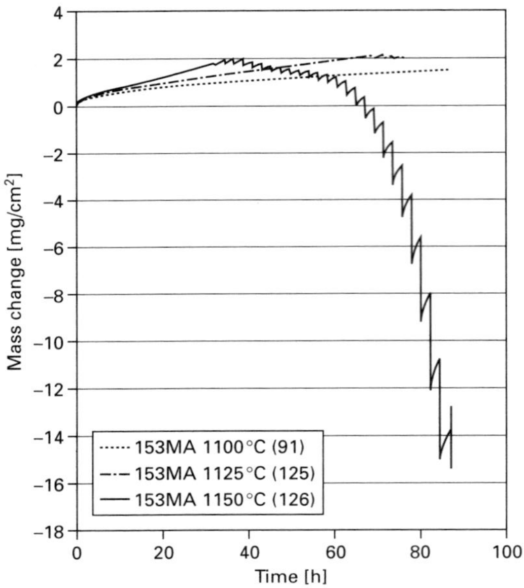

3.1 Austenitic 153MA, 2 h cycle (APAB).

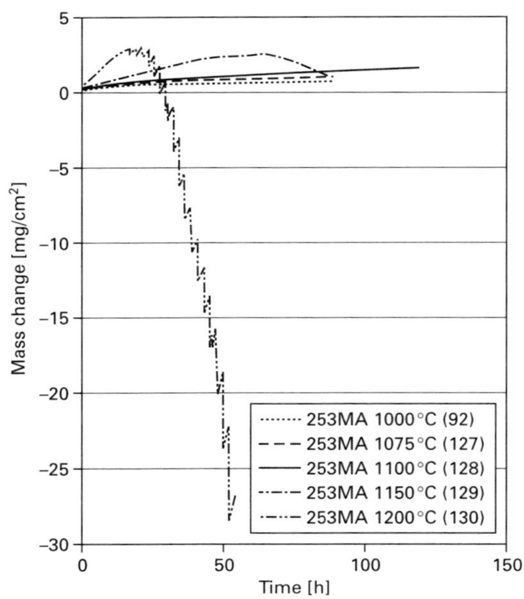

3.2 Austenitic 253MA, 2 h cycle (APAB).

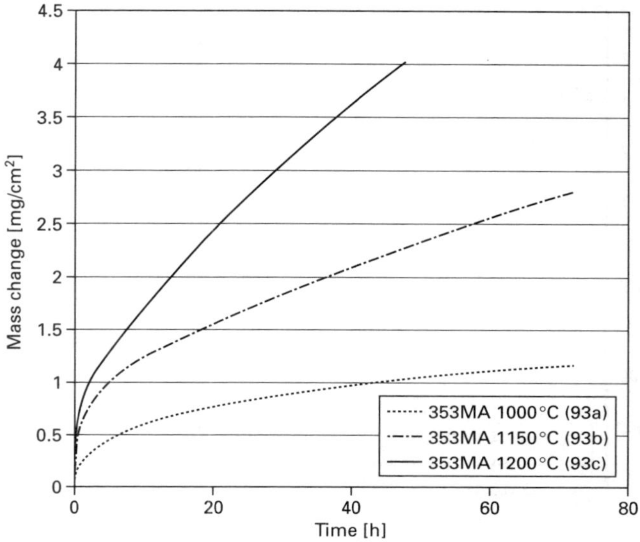

3.3 Austenitic 353MA, 2 h cycle (APAB).

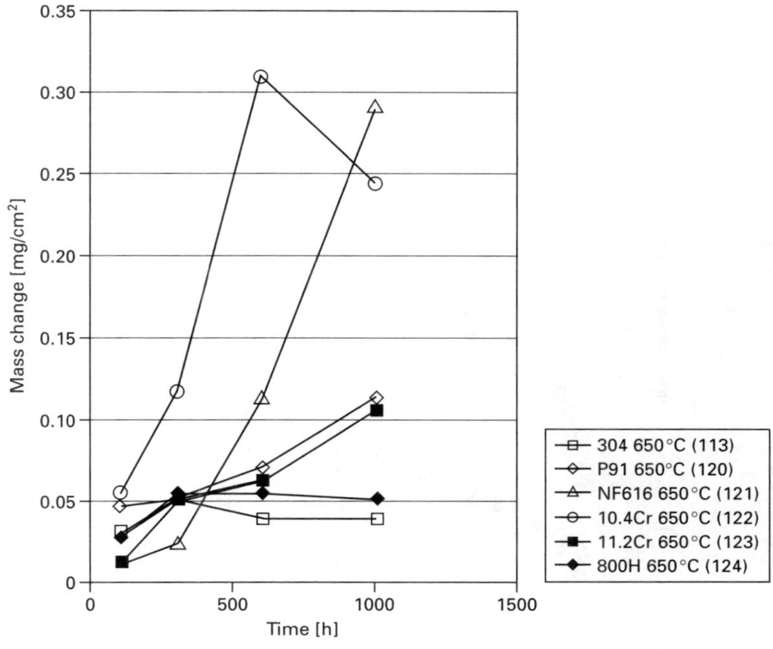

3.4 Austenitic and ferritic alloys, $3^1/_2$ h cycle (KWI-DECHEMA).

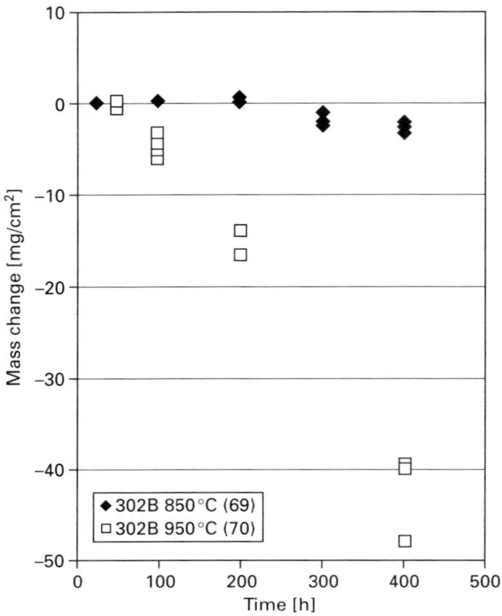

3.5 Austenitic 302B, 20 min cycle (USI).

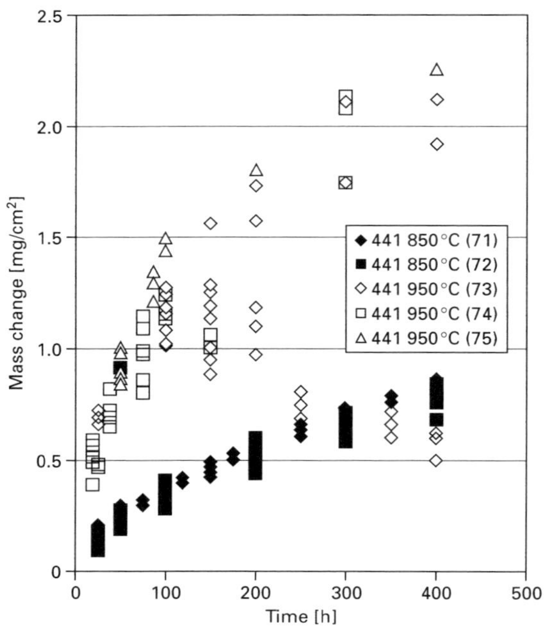

3.6 Ferritic 441, 20 min cycle (USI).

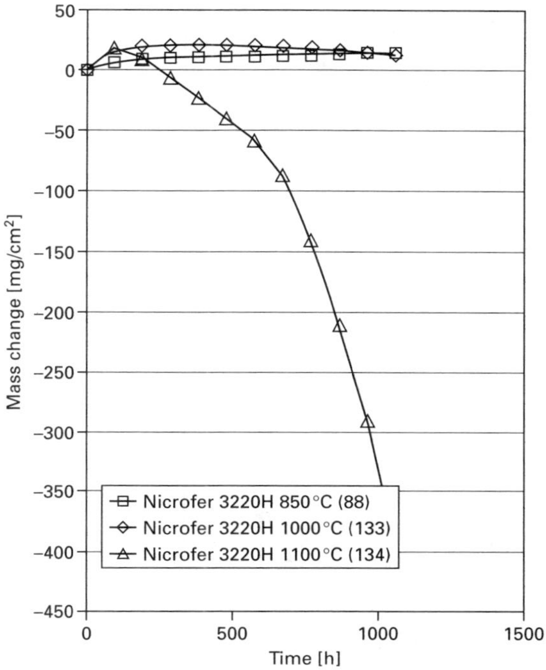

3.7 Austenitic alloy, 96 h cycle (Thyssen Krupp VDM).

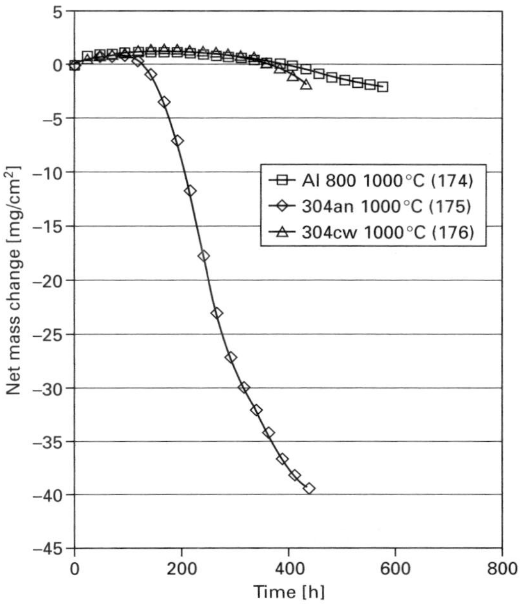

3.8 Austenitic alloys, 24 h cycle (SIMR).

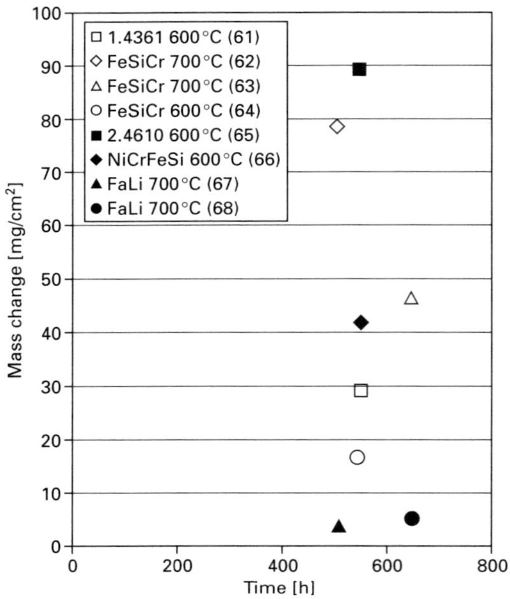

3.9 Austenitic, ferritic FeCrAl and Ni–base alloys, H_2S environment, 24 h cycle (KWI-DECHEMA).

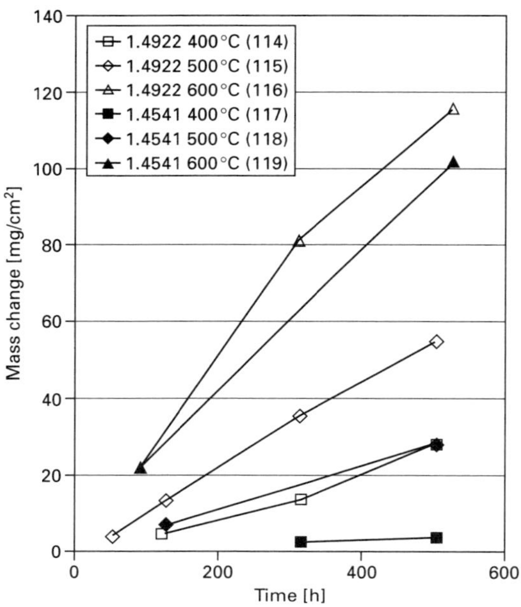

3.10 Ferritic alloys, H_2S environment, 24 h cycle (KWI-DECHEMA).

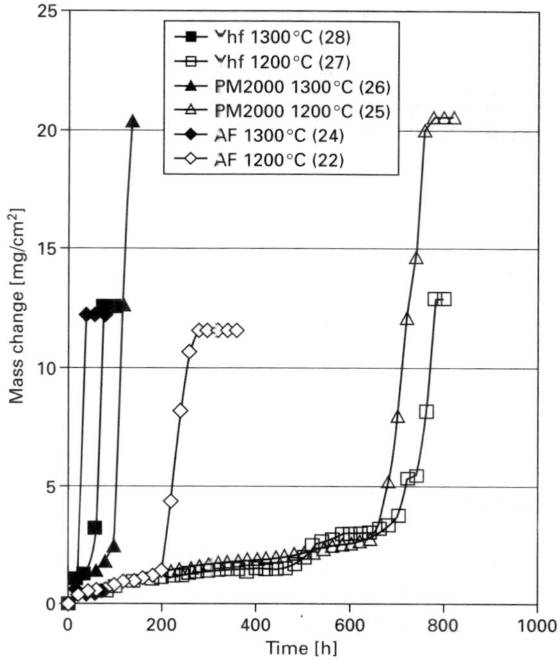

3.11 FeCrAl alloys, 20 h cycle (University of Liverpool).

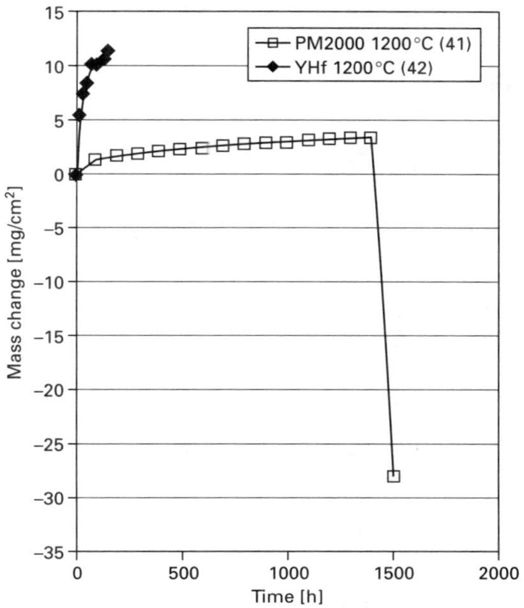

3.12 FeCrAl alloys PM2000 (100 h cycle) and Aluchrom YHf (20 h cycle) (TU-Clausthal).

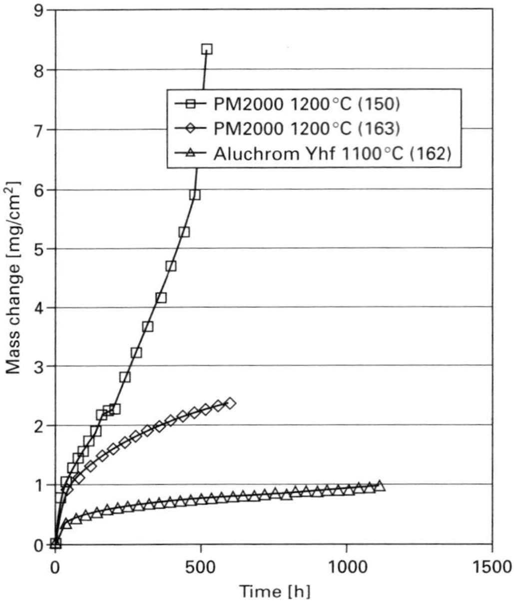

3.13 FeCrAl alloys, 2 h cycle (FZJ).

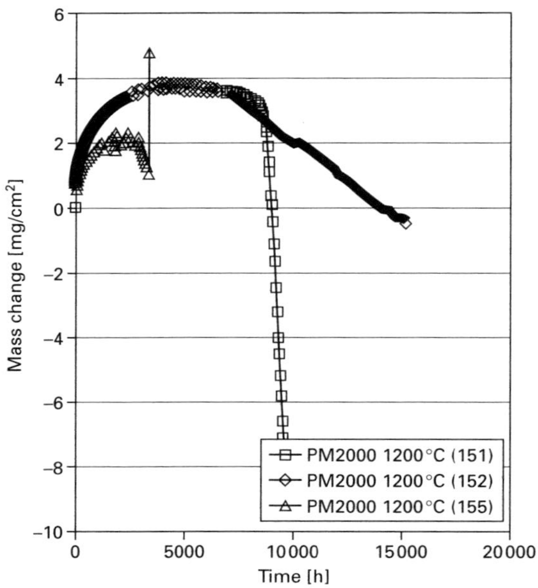

3.14 FeCrAl alloy PM2000, 20 h cycle (FZJ).

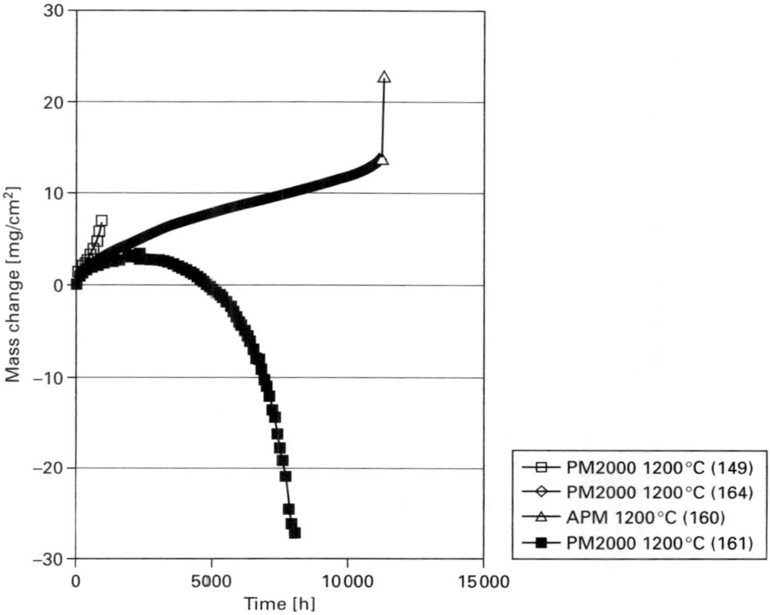

3.15 FeCrAl alloys, 100 h cycle (FZJ).

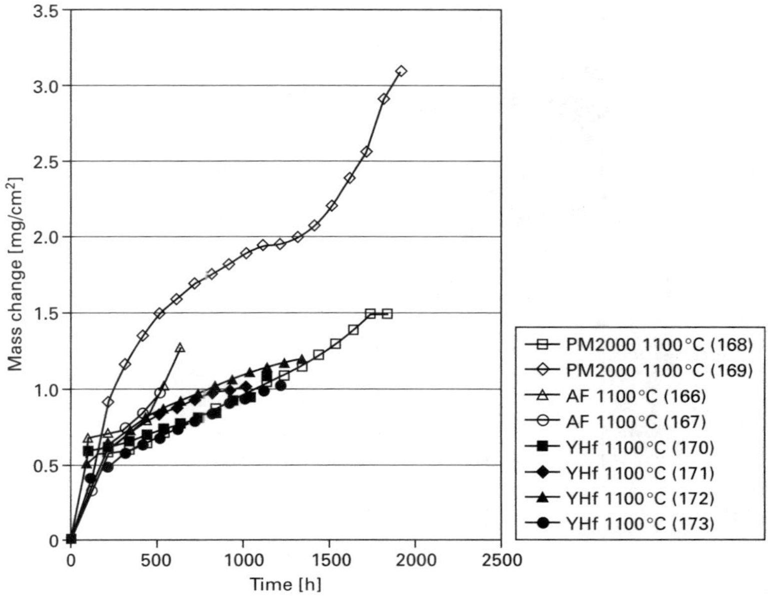

3.16 FeCrAl alloys, 100 h cycle (EC-JRC-IAM).

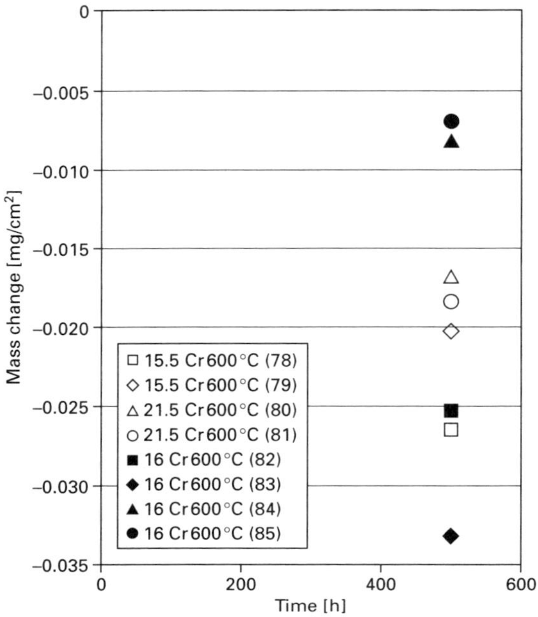

3.17 Ni–base alloys, 1 min cycle, HCl environment (CEA).

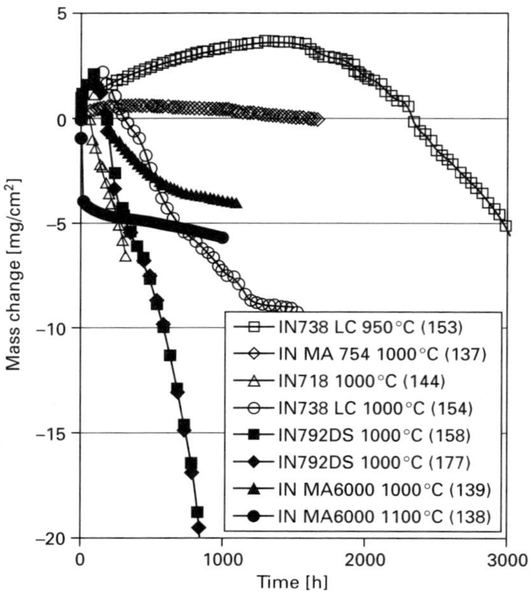

3.18 Ni–base alloys, 2 h cycle (FZJ).

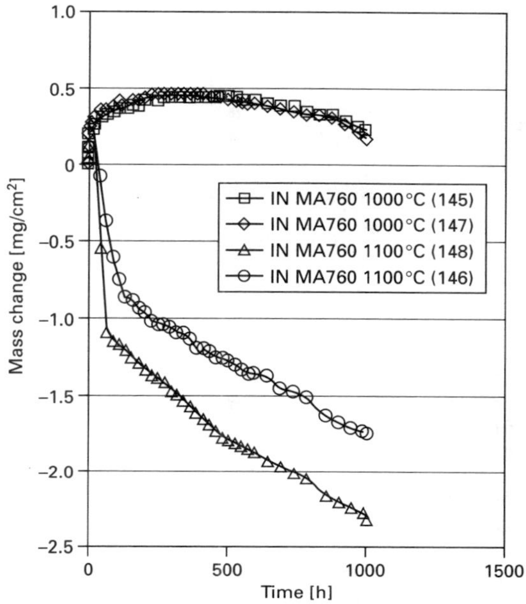

3.19 Ni–base alloys, 2h cycle (FZJ).

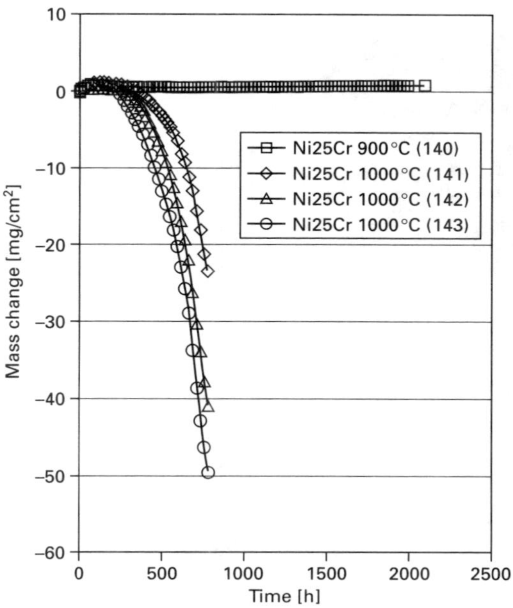

3.20 Ni–base alloys, 2h cycle (FZJ).

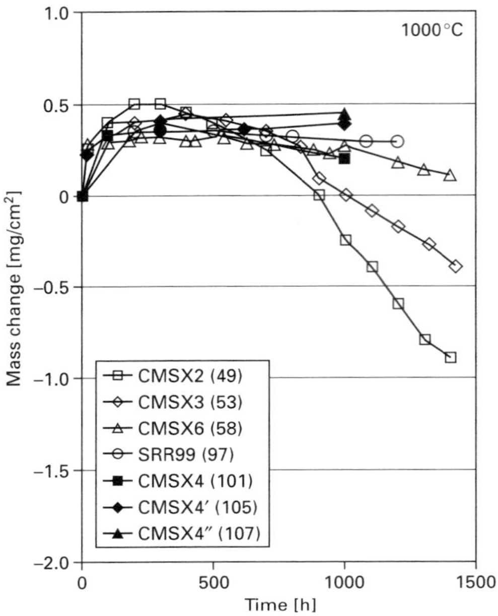

3.21 Ni–base alloys, 1 h cycle, natural cooling (KWI-DECHEMA).

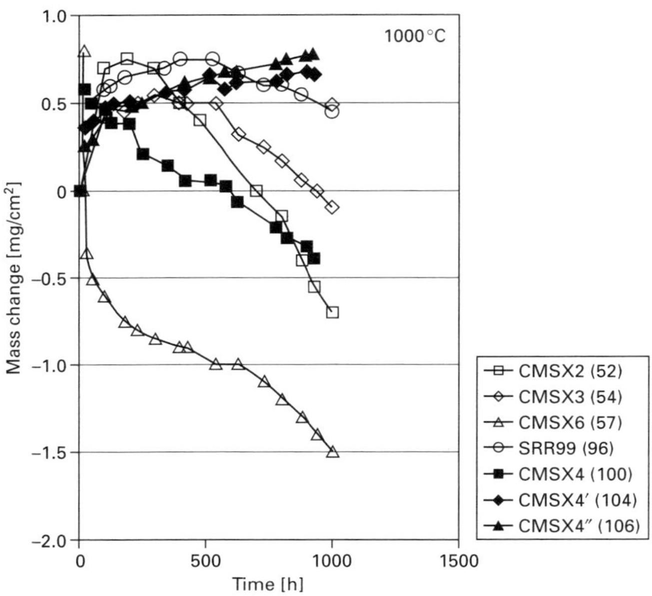

3.22 Ni–base alloys, 1 h cycle, gas blast cooling (KWI-DECHEMA).

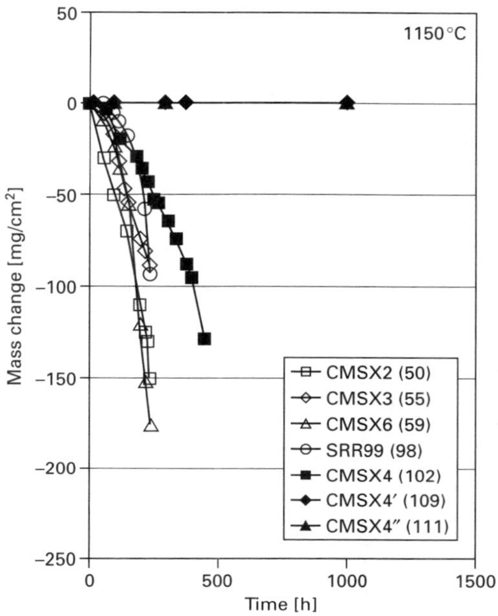

3.23 Ni–base alloys, 1 h cycle, natural cooling (KWI-DECHEMA).

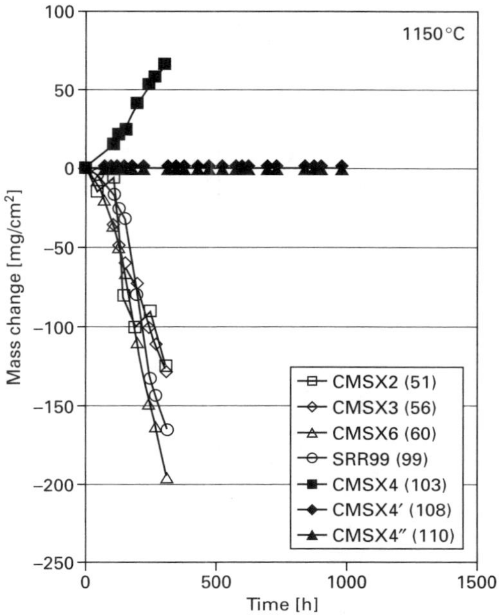

3.24 Ni–base alloys, 1 h cycle, gas blast cooling (KWI-DECHEMA).

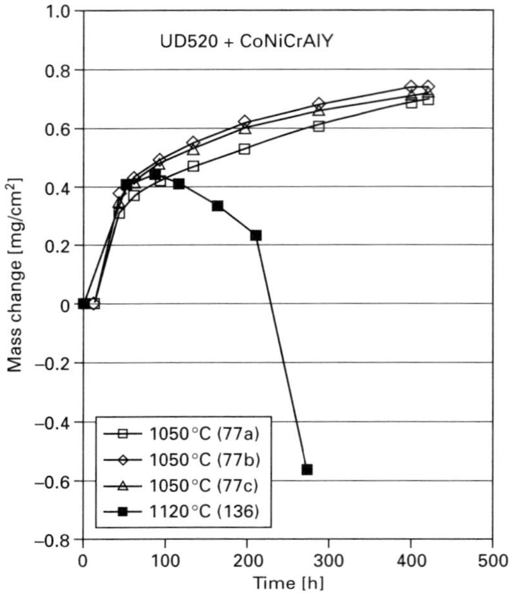

3.25 Coated Ni–base alloy, 1 h cycle (CESI).

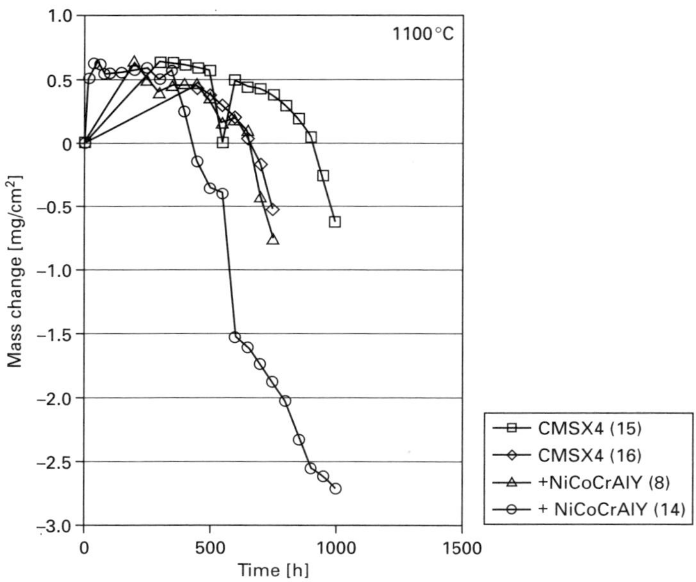

3.26 Coated and uncoated Ni–base alloys, 1 h cycle (DLR).

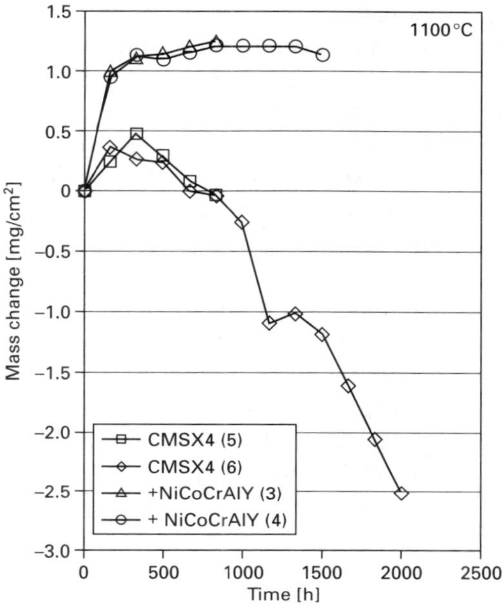

3.27 Coated and uncoated Ni–base alloys, 166 h cycle (DLR).

covariates and correlation analysis have been used to test the relationship between test parameters and output profiles (see, for example, Nicholls and Hancock, 1983). A significance level of 5% is used throughout this chapter.

Figures 3.1–3.3 show very clearly the effect of upper dwell temperature on the output profiles; no further statistical analysis has been carried out. The data in Fig. 3.4 are profiles for both austenitic and ferritic. It is interesting that the profiles – all for the same temperature – differ not in line with the materials group but as a result of some other test parameters not recorded or the combined effects of many sources of experimental variation. Despite the similarity of the profiles in this figure, the alloy groups are usually treated separately.

Figures 3.5 and 3.6 clearly show the effect of upper dwell temperature as does Fig. 3.7. The data in Fig. 3.9 only have a single measurement point and the time of measurement differs. As time of measurement is an important factor in the mass change, neither the effect of upper dwell temperature or group can be evaluated, nor can the experimental variation be reliably assessed.

Figure 3.11 shows profiles that differ in upper dwell temperature and specimen thickness. Thickness coincides with composition and so either factor may be responsible for any difference in the profiles. The data are balanced in that all six combinations of temperature and thickness (composition) are represented. The value of the parabolic rate constant (log (val)) has been provided and the logged value is used in ANOVA to make sure that the assumptions of approximate normality and constant variability are obeyed. Both temperature and thickness (composition) are significant. Figure 3.28 illustrates the effects of the factors. The parallel nature of the lines shows that there is no interaction between the factors.

The standard deviation estimated from the ANOVA residual mean square is entered in Table 3.1. The analysis can be carried out with any other appropriate summary of the output profiles and the experimental variation can be estimated by the standard deviation.

Figures 3.13, 3.14 and 3.15 (IDs 149–152, 155, 160–164) show data that differ in specimen thickness and upper dwell time (related to cycle length). The output profiles do not all show the same parabolic shape and it is not possible to find the time of maximum mass change. As an alternative, all profiles have a value of mass change at 600 h (ID 150 goes to only 500 h) and this is used as a summary parameter. The logged value is suitable to be used in ANOVA. All except ID 162 have upper dwell temperature of

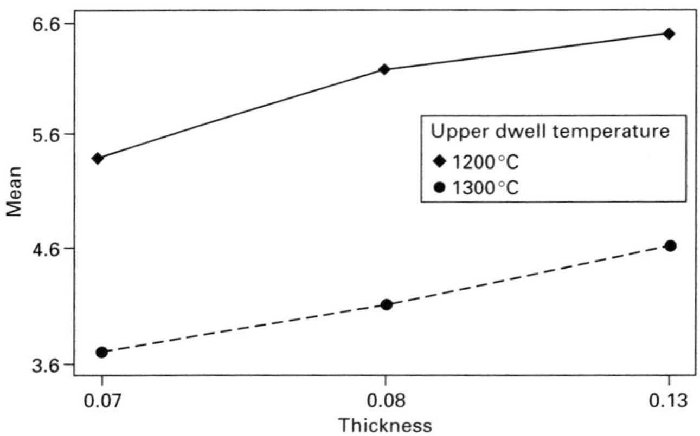

3.28 Interaction plot – log mass change vs thickness and upper dwell temperature.

Table 3.1 Estimated experimental variation in terms of standard deviation

ID	Figure	Output parameter	Standard deviation
22, 24–28	3.11	Log(rate)	0.13
149, 160–1, 164	3.15	Log(mass change at 600 h)	0.12
166–173	3.16	Mass change at 500 h	0.26
170–173	3.16	Mass change at 500 h	0.10
78–85	3.17	Mass change at 500 h	0.009
145–148	3.19	Log(T_{max})	0.09
141–143	3.20	Mass change at 100 h	0.22
141–143	3.20	Time at max mass change	26.6
49–60, 96–111	3.21–3.24	Mass change at 150 h	0.012

1200 °C. The log of mass change at 600 h increases with thickness (not shown) but neither this effect nor upper dwell time (Fig. 3.29) is significant. This may be because 600 h is too early to pick up the differences.

The data in Fig. 3.16 (IDs 166–173) shows profiles for different alloys within the FeCrAl group. The profiles for ID 166 and 169 seem to be different from the rest; however, none of the test parameters differs for these two alone. The profiles do not reach a maximum so t_{max} cannot be found, instead the output parameter of mass change at 500 h is used to summarise the data. The experimental variation is estimated by the standard deviation as 0.26 mg/cm^2. Data from ID 170–173 are all from the same designation and they are more consistent. An estimate of standard deviation (SD) for these four IDs is 0.10 mg/cm^2.

The data in Fig. 3.17 (IDs 78–85) all have a single value at 500 h. The data differ in composition and also in that IDs 82–85 were pre-oxidised. Neither of these test parameters has a significant effect on the mass change at 500 h. The experimental variation is estimated at 0.009 mg/cm^2.

The profiles in Fig. 3.18 show the effect of upper dwell temperature in that the time of maximum mass change is later for ID 153 and earlier for ID 138. Within the data in this figure, there are six results with the same upper dwell temperature of 1000 °C and these differ in composition and surface area. Disregarding composition and using the mass change at 100 h, the effect of surface area is significant with smaller mass change for larger surface area.

The data in Fig. 3.19 (IDs 145–148) clearly show the effect of upper dwell temperature but also provide an opportunity to look at experimental variation. The two upper profiles for IDs 145 and 147 are at 1000 °C and show a maximum mass change at t_{max} = 277 and 231 h respectively. The two lower profiles for IDs 146 and 148 are at 1100 °C and both show a maximum

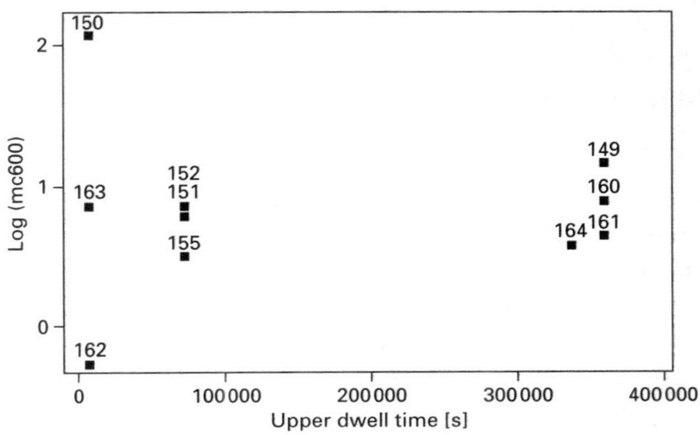

3.29 Scatter plot – log mass change vs upper dwell time.

mass change at 12h. Although there are only four pieces of information, ANOVA of log(t_{max}) gives an estimate of experimental standard deviation of 0.09 with 2 degrees of freedom.

Figure 3.20 clearly shows the effect of temperature on the output profiles. The data in Fig. 3.20 for IDs 141–143 all differ in length and width of specimen and designation, and so the effect of these test parameters cannot be investigated. IDs 141 and 143 have an end test date a month after ID 142. The times at maximum mass change are 185, 185 and 139 respectively. These can be used to estimate experimental variation at 1000 °C; it is 26.6h with 2 degrees of freedom. Using mass change at 100h, the data are 0.79, 0.83 and 1.19 mg/cm^2, giving a standard deviation of 0.22 mg/cm^2.

Figures 3.21 to 3.24 show the relationship of upper dwell temperature and type of cooling with output profiles. It can be seen that the gas blast cooling increases the variability of the responses compared with natural cooling. The increase in temperature causes the profiles to miss the first stage of oxidation. The profiles are difficult to compare statistically. Using mass change at 150h, the standard deviation after allowing for upper dwell temperature and type of cooling is 0.012 mg/cm^2.

Figure 3.26 shows two profiles from coated Ni–base alloys and two from uncoated. There are too few tests for the effect of coating to show up as significant in a statistical test although the effect is indicated on the figure.

3.4 Statistical analysis – across data sources

3.4.1 Long dwell data

There are 25 results for FeCrAl, of which 21 have complete data to 100h and 23 to 500h. The test parameters that vary and have complete data are upper dwell temperature and time, thickness (and width and length), method of cooling and environment. The data are incomplete for many of the other test parameters, which differ including lower dwell temperature and time, position of recording, type of thermocouple, temperature stability, heating time, cooling time, production route, heat treatment, ambient temperature, etc. Most of the test parameters vary in step with source, however, and so their effects are not distinguishable from each other.

Working with mass change at 100h, the sources differ significantly. In the light of the above analysis of data within sources, this is probably because the sources mostly have different upper dwell temperatures. The mass change at 100h increases with upper dwell temperature. Specimen thickness varies across the data sets as do surface area, width and length. Only thickness has a significant effect, with mass change increasing with specimen thickness. The effect of upper dwell time is also significant with mass change increasing with upper dwell time. The remaining experimental variation is estimated by the ANOVA standard deviation at 0.14 mg/cm^2.

Mass change at 500 h is correlated with mass change at 100 h as expected but using mass change at 500 h only upper dwell temperature now shows a significant effect because of the increased variation in the data. The experimental variation is estimated by the ANOVA standard deviation of 0.30 mg/cm^2 which is larger than that in Table 3.2 for Fig. 3.16 as expected.

There are nine results for Ni–base from three different sources including three different upper dwell temperatures: IDs 133 and 134 are at 1000 and 1100 °C, IDs 3–6 are at 1100 °C and IDs 88 a,b&c are at 850, 1000 and 1100 °C. Heating time, cooling time and specimen shape vary in step with each other and cannot be distinguished. Composition also varies but is correlated to source. The relationship of t_{max} and upper dwell temperature is significant with t_{max} decreasing with increasing temperature. The difference between sources is not significant which is good for future experiments. The experimental variation is estimated by the ANOVA standard deviation of 265 h, which is larger than that for the short dwell data in Table 3.1 for Fig. 3.20 as expected.

3.4.2 Short dwell data

There are 46 results for Ni–base but the scales differ markedly. For example, in Fig. 3.22, ID 96 has mass change values less than 1 mg/cm^2 and in Fig. 3.24, ID 99 has values over 100 mg/cm^2. Analysis of these data shows up the effect of upper dwell temperature above all. The standard deviation may not be very reliable.

There are nine results for ferritic that have data on test parameters that can be used. Heating time varies, but is in line with upper dwell temperature. Using the mass change at 100 h, upper dwell temperature is a significant factor. The remaining experimental variation is 0.05 mg/cm^2.

3.4.3 Ultra-short dwell data

The ultra-short Ni–base data all come from the same source and have been analysed in Section 3.3 (Fig. 3.17).

3.4.4 Combined analysis

In principle all the data could be combined in an overall analysis which would be powerful because of the large numbers involved. The problem with that approach is that it would not be possible to find a common output summary parameter to suit all the output profiles.

3.5 Number of replicates required for future experiments

Work Package 5 refers to planning for future experiments. It was suggested to have three dwell times, three cooling rates, two cooling times and three upper dwell temperatures. One suggestion was to have 42 samples, another to have only nine samples. One efficient way of experimenting with several factors (test parameters) is to have a balanced experimental design where each factor level is tested an equal number of times. Formulae exist for determining the number of samples needed for an experiment. The number of samples can also be found in statistical software.

The following rule-of-thumb can be used to give an idea of the power of an experiment depending on how many samples are used. In the equation:

$$n > 20\,\frac{(\mathrm{SD})^2}{(\mathrm{difference})^2} \qquad 3.1$$

n is half the total number of samples for a 2-level factor and a third of the total number of samples for a 3-level factor, the SD is the standard deviation, which can be estimated from ANOVA, and 'difference' is the size of true difference between mean values at different factor levels that is likely to be detected as significant.

It can be seen from the equation that the larger the value of SD, the larger the number of samples needed to detect a particular difference. For a particular SD, the larger the number of samples, the smaller the difference which can be detected as significant. Therefore the larger the value of n, the more sensitive the experiment becomes:

$$\mathrm{difference} > \sqrt{20}\,\frac{(\mathrm{SD})}{\sqrt{n}}$$

It can be seen that an experiment with ten samples at each factor level will only be powerful enough to show up a true difference as large as $\sqrt{(20/10)}$ = 1.41 times the SD. So for example, using the SD in Table 3.1 for t_{max} in Fig. 3.20 and a cooling rate factor, the difference between t_{max} at one cooling rate and t_{max} at the other cooling rate must be greater than $1.41 \times 26.6 =$ 38 h for cooling rate to be detected as a significant factor.

Future experiments will be carried out at several locations (sources) and so the most appropriate estimates of SD are from the analysis of data across sources. These SDs will typically be larger and so indicate the need for more trials.

If none of the factors being tested is thought to have an effect as big as that calculated by equation 3.1 then more samples are needed. Another option is to change the levels of the factors to give a wider range: for example, a wider range of cooling rates may make the effect of cooling rate more obvious but the levels should still be reasonable so that the experiment is representative of normal practice.

If the aim of the experiment is to find mean values at different factor levels, then it must be remembered that the confidence limits for mean values depend on the sample size. In general a 95% confidence interval is approximately mean $\pm 2 \times \mathrm{SD}/\sqrt{n}$.

3.6 Conclusions

- The database contains a lot of interesting data.
- Values for some key test parameters are missing.
- The output measurements vary considerably in their number and frequency making it difficult to summarise them consistently.
- The usefulness of the data is limited by the unbalanced way in which the test parameters vary making fair comparisons difficult. Future experiments should be planned so that full assessments of the data can be made.
- The database is now in a well-organised format so that further analysis of different combinations of results can be carried out as required.

3.7 Recommendations

Modelling software, such as winCOSP (Smialek and Auping, 2002) could be used to summarise the mass change profiles in future. The winCOSP authors acknowledge that complex behaviour and irregularities are common and may make the models unable to fit the data.

A standard number of measurements would be useful. Examples in Smialek and Auping (2002) have 1 h cycles and 9, 18 or 50 measurements from which models are fitted.

Experiments should be planned so that the maximum information can be obtained from the minimum number of trials. This means deciding on test parameters of interest and making sure that there are sufficient unbiased trials at each level of the test parameters so that their effects can be assessed.

3.8 References

J.R. Nicholls and M.J. Bennett (2000) 'Cyclic oxidation – guidelines for test standardisation, aimed at the assessment of service behaviour', *Materials at High Temperatures*, **17**(3), pp. 413–428.

J.R. Nicholls and P. Hancock (1983) 'The analysis of oxidation and hot corrosion data – a statistical approach', *High Temperature Corrosion, NACE-6*, pp. 198–210.

J.L. Smialek and J.V. Auping (2002) 'COSP for windows: strategies for rapid analyses of cyclic oxidation behaviour', *Oxidation of Metals*, **57**, pp. 559–581.

Part II

Experimental investigations on the influence of test parameter variation on thermal cycling oxidation behaviour

4

Standardised test procedures, definitions and statistical design of experiments for investigation of test parameter variation on thermal cycling oxidation testing

M. SCHÜTZE and M. MALESSA, DECHEMA e.V., Germany and S. COLEMAN, Newcastle University, UK

4.1 Introduction

In industrial applications, metallic materials are subjected to different types of thermal cycling. Therefore it would be counter-productive to develop one single standard for the cyclic oxidation testing of all types of materials in all industrial applications. Hence three different sets of test procedures applicable to a range of service conditions found in a large number of industrial applications have been developed.

The three different test types *(long dwell times, short dwell times and ultra-short dwell times)* have been selected in such a way that together they cover practically the entire range of service conditions in which high-temperature materials are subjected to thermal cycling. In order to limit the number of parameters and to facilitate the data comparison, preparation of corrosion test specimen, test procedures and data analysis are proposed following whenever possible the recommendations prepared within the European project TESTCORR that aimed to develop standardised test procedures for isothermal testing [1].

4.2 Preparation of corrosion test specimen and equipment

The results and the success of the testing depend on several key factors defined in TESTCORR and listed below. In this project, in order to limit the number of parameters, some of them have been fixed while the influence of others has been investigated.

4.2.1 Preparation of corrosion specimen

Specimen size and shape used within the COTEST project

- For long and short dwell times: block : $12 \times 17 \times 1.5\,mm^3$

• For ultra-short dwell times:	wire diameter 0.4 mm × 150 mm wire diameter 0.7 mm × 150 mm foil 50 μm thick × 5 mm × 150 mm
• For burner rigs	rod diameter 8 mm × 120 mm rod diameter 6 mm × 50 mm
• For test under deposit	Rods diameter 6 mm × 50 mm long or block 50 × 6 × 1.5 mm^3

Grain size and structure

To be characterised for each grade.

Specimen manufacturing route

To be defined by each metal supplier and to be reported.

Final surface condition and cleaning of specimens

The final surface condition has a strong effect on the results of testing. In order to avoid any surface condition variabilities due to, for example, the different specimen preparation or industrial surface finish, the following procedure was used for the specimens to be tested in the frame of this project.

For long and short dwell times:

- Grinding grade 1200 under water (different grade used, grinding time and pressure have to be evaluated by each participant). Edges also have to be ground. No sharp edges must be present.
- Ultrasonic cleaning method in an alcohol bath for minimum 10 minutes and flushing in an additional bath of 'Analar' isopropanol after which the specimens have to be hot air dried, visually inspected and placed overnight in a desiccator before weighing (between 12 and 48 h). A new bath has to be used for each new series.

For ultra-short dwell times:

- The as-manufactured form is subjected to ultrasonic cleaning as described above.

Specimen identification

No imprinting of the specimen, which could disturb the oxidation behaviour of the metal but the use of precise documented boxes and papers.

Time between end of cleaning and beginning of the test

TESTCORR suggests a time of 12–48 h (overnight in desiccator after cleaning). It has to be reported by each participant.

Weighing of specimen

A minimum of three replicate weighings following the TESTCORR recommendations. The accuracy of the balance must be defined.

Reproducibility

A minimum of three specimens per testing condition shall be used.

4.2.2 Preparation of equipment

Gas atmospheres, composition, deposits, flow rate

Investigate parameter variation.

Thermocouple

- Type: type S (Pt–10% Rh/Pt) or R (Pt–13% Rh/Pt) should be used.
- Calibration: at least every 6 months.

Furnace

- Calibration: as per ASTM E1350-91 see 'TESTCORR'. This should be undertaken before each new set of tests with a thermocouple attached to a dummy specimen.
- Temperature control: two thermocouples located near to the specimens.

For ultra-short testing, pyrometers will have to be used both for temperature control and to allow heating and cooling rate calculation avoiding the 'load' of the specimens with the heat capacity of the thermocouple. Pyrometers will be calibrated against a thermocoupled reference of each alloy under test, as per furnace calibration.

Definition of the dwell time

The effect of non-isothermal oxidation has been calculated with the help of mathematical software (see Chapter 5). For the alloy Incoloy MA956 comparative measurements have been carried out. The influence of the heating period of the sample on total oxide growth has been investigated and compared with the isothermal oxidation during usual oxidation experiments or the dwell time in cyclic oxidation.

Following these results, the recommendation is that the dwell time begins at 3% under the upper temperature and finishes at removal of the specimen from the hot zone.

Calculation of the heating and cooling rates

Heating

- Slope of the temperature increase between the introduction of the specimens in the furnace and the start of the dwell time at the upper temperature (average value).
- Fastest slope of the temperature increase.

Cooling

- Initial slope of the temperature decrease.
- Time between the removal of the specimen from the furnace until the specimen reaches 50 °C

Time at lower temperature

The cooling time lasts from the removal of the specimen from the furnace until the specimen reaches 50 °C. Then the specimen will stay for additional time at this temperature or below. The time at low temperature is 15 or 60 min for the short dwell tests.

In the case of long dwell tests the introduction of an additional cooling time would disturb the 24 h operation rhythm. This cooling time can be subtracted from the cold dwell times (under the precondition that the specimens have been cooled to at least to 50 °C) to maintain the 24 h rhythm. In any case, the real thermal cycle has to be recorded by each participant.

4.3 Different thermal cycles investigated

The following three types of tests will be addressed.

4.3.1 Long dwell times

This type of testing aims to simulate conditions in large-scale industrial facilities encountered in applications such as power generation plants, the chemical industry, waste incineration plants and the process industry. In these applications the metallic components are designed for extremely long term operation, e.g. typically 100 000 h. Thermal cycling of materials occurs due to planned plant shut-downs, e.g. for regular maintenance, or due to unplanned shut-downs as a result of offset conditions. Therefore, the time intervals between various thermal cycles are relatively long and the number of cycles is, related to the long operation time of the components, quite small, i.e. typically around 50.

- *Adopted thermal cycle*: Investigation of thermal cycles with 4, 8 and 20 h hot dwell time and 2 and 4 h of cooling and reheating (including handling time for specimen weighing).
- *Data measurement (mass change) frequency*: once a working day.
- *Total exposure time*: 300 hours

4.3.2 Short dwell times

This type of thermal cycling is typically experienced in applications such as industrial gas turbines, aero engines, heat treatment facilities and furnaces. The intervals between start and shut-down of the facilities are generally much shorter than in applications described above. Also, the design life and/or the time until complete overhaul/repair (typically 10 000–30 000 h) are much shorter and, depending on the specific practical application, the number of cycles is much higher than in the case of long dwell exposure.

- *Adopted thermal cycle:* Different cycles will be studied with a 0.5, 1 and 2 h dwell time and 1 and 0.25 h cooling and reheating.
- *Data measurement (mass change) frequency*: Taking into account the influences due to frequent interruptions for weight measurements (= extended cold dwell), the reduced number of weight measurements to 7 to 10 (7 is the recommend minimum according to TESTCORR) with more measurements in the early phase of testing (quasi-logarithmic scale) will be realised.
- *Total exposure time:* 100 h originally planned, but extended to 300 h.

4.3.3 Ultra-short dwell times

This type of testing mainly addresses applications of high-temperature alloys as heating elements in the form of wires or foils. Another typical application in which such short cycles prevail would be catalyst foil carriers, e.g. in cars. In such applications the number of cycles is extremely high in relation to the overall design life (typically several hundred to a few thousand hours), and the time intervals between heating and cooling can be as low as minutes or even seconds. Such conditions are commonly encountered in a number of industrial applications such as, for example, burners and hot gas filters but also in a large variety of domestic applications where metallic heating elements are used, e.g. in cooking plates, toasters, boilers, dryers, fryers.

These ultra-short testings can be achieved in two ways: by traditional Joule effect heating (normal procedure used by industries producing electrical materials (limited to conducting materials) or by light (lamps or laser) heating for non-conducting materials. A review of alternative methods has been undertaken and a design for a focused light-heated furnace proposed. Following discussions with industrial partners (Krupp VDM, Kanthal), it has been decided to adopt the Joule heating method, based on a modification of existing ASTM standards. Cranfield University has the resources to undertake the design of modified Joule heating test facilities that incorporate environmental testing. Experiments are required in Work Package 5 to evaluate this facility, and to generate data comparing resistance measurement to mass change.

- *Adopted thermal cycle*: Different cycles will be studied with 2, 5, 10 min dwell times and 5 min cooling and reheating.
- *Data measurement/frequency*
  - Resistance – continuously monitored.
  - Mass change – samples will be removed after a range of cycle/exposure times to determine mass gain and correlate this with resistance change. Samples will be removed after 10, 20, 40 and 100 h of testing.
- *Total exposure time*: 100 h (or early failure).

4.4 Analysis of results and post-test evaluation

4.4.1 Data management

The following parameters had to be reported:

- Test report
- Name of the test
- Identification of the test rig
- Specimen: material, geometry, sizes, identification
- Surface preparation and cleaning procedure
- Testing atmosphere composition
- Gas flow rate and pressure
- Deposit (if applicable)
- Time between end of cleaning and beginning of testing
- Start time
- Stop time
- Thermal cycle (temperature, heating and cooling rates, dwell time, method of cooling)
- Number of cycles
- Intermediate steps (if applicable)
- Notes
- Results.

4.4.2 Evaluation of corrosion rate constants

The mass change measurements have to follow the TESTCORR recommendations to obtain the net mass change of specimens and the gross mass change. The results should be expressed as the mean value of mass ± 2 standard deviations.

4.4.3 Data post-evaluation (metal loss, cross-section measurements)

Data post-evaluation can be obtained either by metal loss after descaling or by cross-section measurements. It should be preferably based on material recession or scale thickness measurement (TESTCORR) by preparing a metallographic cross-section following TESTCORR recommendations or Simms *et al.* [2]. Care has to be taken in mounting the specimen normal to the primary axis of the specimen.

These measurements use usual light microscope without additional preparation and will be done on all specimens.

Measurement shall consist of:

- deposit thickness
- scale thickness
- depth of internal penetration
- depth of grain boundary attack measured
- depth of any depleted zone
- remaining unaffected material.

A minimum of 24 measurements per specimen shall be obtained. The results should be expressed as the mean value of mass ± 2 standard deviations. In addition, the position of maximum attack should also be measured and reported.

More detailed characterisation of corrosion product composition for selected specimens will be accomplished by scanning electron microscopy, energy dispersive X-ray analysis and X-ray diffraction according to proposal.

4.5 Statistical design of experiments

4.5.1 Experimental design for investigation of selected materials under cyclic oxidation conditions

All investigations were carried out in experiments designed according to statistical criteria. The aim was to ensure that the results of the experimental trials could be analysed statistically, to test the effects of all factors in a balanced, robust way. Designs varied for the different materials and circumstances in which they were to be tested. There were specific experimental plans for long, short and ultra-short dwell experiments and for complex environments.

There are many experimental designs and they are described in statistical texts such as Grove and Davis [3]. For long and short dwell experiments, the main linear and quadratic effects of upper dwell temperature and upper dwell (hot) time and the main effects of lower dwell time (cold time) and environment were thought to be of interest. A quarter fractional nine trial design was chosen.

An 18-trial experiment was designed for ultra-short dwell tests according to statistical criteria to investigate the effects of four parameters. In cases where fewer trials could be carried out, designs were constructed to provide a balanced comparison of the effects of various cycling parameters. The balanced design enables:

- a prediction equation to be made that interpolates between the readings;
- significant test parameters to be determined;
- a confidence interval for the predictions to be obtained.

The number of test parameters and levels being investigated in only nine trials:

- may not allow us to investigate interactions (synergies) between test parameters;
- and if there are strong synergies these may bias the effects of some of the test parameters.

The test matrices of the experiments are given in the following chapters.

4.6 References

1. Code of Practice for Discontinuous Corrosion Testing in High Temperature Gaseous Atmospheres, TESTCORR project CEC contract no. SMT4-CT95-2001.
2. N.J. Simms, J.E. Oakey, J.R. Nicholls (2000) 'A Methodology for the Quantification of Corrosion and Erosion Damage in Laboratory, Burner Rig and Plant Environment, *Electrochemical Society Proceedings* 99, 305–316.
3. D. Grove and T. Davis (1992) *Engineering, Quality and Experimental Design*. John Wiley and Sons, New York.

5
The effect of heating on the total oxidation time

G. STREHL and G. BORCHARDT,
Schmidt+Clemens GmbH & Co. KG, Germany

5.1 Introduction

In the framework of the COTEST project, experimental cyclic oxidation parameters are investigated that seem to be crucial for scatter in the results between different laboratories. The standard to be developed shall cover a temperature range from 500 to 1500 °C. Heating rates vary from slow furnace heating, 0.17 K/s to ultra-fast heating in a burner rig, 6.6 K/s. Cooling rates are determined by slow cooling in a furnace or by forced cooling (cold air blast) and cover the same range.

During the discussion the question was raised, which period of time should be defined as dwell time at high temperature? Which role does the heating period play, especially when the cycles are very short? In this chapter some calculations are collected, which should give some guidance in answering these questions. The experimental part refers to the alloy Incoloy MA956, which is very well known as an alumina-forming high-temperature material.

5.2 Heating of the sample

When starting an oxidation experiment or a cycle in cyclic oxidation the samples are usually introduced into the furnace and exposed directly to temperature. There are three modes of heat transport involved in heating the sample: heat conduction through the gas in the furnace, convective heating of the sample surface, where again the gas is the transport medium and heating by radiative transport. Because of the big temperature difference between the sample and the furnace convection in the gas will immediately start, thus heating by conduction can be neglected.

5.2.1 Radiative heat transport

The radiative heat transport can be very important, certainly at temperatures above 800 °C. This calculation is based on the Stefan–Boltzmann law, describing the excitation M, the power emitted by a region of surface divided by the area of the surface,

$$M = \varepsilon \sigma T^4 \qquad 5.1$$

where σ is the Stefan–Boltzmann constant ($\sigma = (2\pi^5 k^4)/(15c^2h^3) = 5.67051 \times 10^{-8}\ W/(m^2K^4)$) In this calculation the furnace and the sample are treated as grey radiation sources, so they can be described by their emission coefficient ε only. Because of Kirchhoff's law the absorption coefficient a is equal to the emission coefficient ($a = \varepsilon$). The thickness of the material leads to zero transparency ($\tau = 0$). For energy conservation reasons the reflection coefficient is $r = 1 - \varepsilon$.

If the sample is treated as a small, cold body with a convex surface A_s, inside a concave volume with a hot surface A_f, which shall represent the furnace, the heat flux $\dot{Q}$ from A_f to A_s can be calculated, using standard formulae, as given in, e.g., Baehr and Stephan [1]

$$\dot{Q} = \varepsilon_{fs} A_f \sigma (T_j^4 - T_s^4) \qquad 5.2$$

For this arrangement the radiation exchange number (f_s) is given by:

$$\frac{1}{\varepsilon_{fs}} = \frac{1}{\varepsilon_f} - 1 + \frac{A_f}{A_s}\frac{1}{\varepsilon_s} \qquad 5.3$$

The surface of our standard rectangular sample (20 mm × 10 mm × 1 mm) is $A_s = 460\,\text{mm}^2$. This would be equal to a disc-shaped sample with a thickness of 1 mm and a diameter of 8.071 mm. In Saunders and Nicholls [2] a sample surface of $500\,\text{mm}^2$ is recommended. The surrounding radiating surface of the furnace can be approximated by a tube with an inner diameter of d = 40 mm and length b:

$$A_f = 2\pi \frac{d}{2} b \qquad 5.4$$

The length b can be determined either by the length of the hot zone of the furnace or by consideration of the radiative heat flux from each surface element of the tube to the sample. The latter gave the approximation $b = \sqrt{2}d$. Using this result, the length of the radiating furnace surface is $b = \sqrt{2} \times 40\,\text{mm} = 56.6\,\text{mm}$, which is in close agreement with hot zones estimated from thermoprofiles in existing furnaces. The radiating area of a furnace tube can be calculated an:

$$A_f = \sqrt{2}\pi d^2 \qquad 5.5$$

For d = 40 mm this gives $A_f \cong 7000\,mm^2$. Thus the radiating surface is 14 times larger than the sample surface and by a factor $\sqrt{2}$ larger than the surface of a sphere with the same diameter as the tube.

The inner lining of most furnaces is made from ceramic materials. As shown above, the radiating surface is much larger than the surface of the sample. For these two reasons the furnace can be approximated by a black body. For the calculation $\varepsilon_f = 0.9$ is used.

The sample is often of a metallic nature and covered with an oxide layer. For metals the emission coefficients vary from 0.018 (polished silver) via 0.09 (polished nickel) to 0.2 (polished iron). During the oxidation process the emission coefficient increases dramatically, e.g. for copper from 0.03 to 0.45 or for aluminium from 0.04 to 0.25. (All data were taken from refs [1] and [3].) For the calculation we use a medium value of $\varepsilon_s = 0.15$.

Taking all these estimations together the heat flux from the furnace to the sample by radiation can be calculated using equation (5.2) with $\varepsilon_{fs} = 0.011$ and $A_f = 7000\,mm^2$.

5.2.2 Convective heating with constant thermal transition coefficient

The heat transport through the gas is described by convective heating, because, as stated above, the big temperature difference between the sample and the furnace immediately leads to a gas flux, thus the thermal conduction through the gas can be neglected. For the calculation of the convective heating two approximations are widely used.

The first is given by a simplified version of Fourier's law:

$$\dot{Q} = A_s \dot{q} = A_s \alpha (T_f - T_s) \qquad 5.6$$

where α is the thermal transition coefficient. T_s is again the temperature of the sample. Values for α differ a lot, but 20 W/(m^{2}K) [4] seems to be a good mean value. The thermal transition coefficient can be understood as the thermal conductivity of the gas divided by the thickness of the gas flow layer along the surface of the sample. With the thermal conductivity of air at 800 °C λ = 0.07154 W/(m/K) [3] the thickness of the gas flow layer δ = 3.6 mm, which fits quite well into the picture.

The second approximation is based on a solution of the partial differential equations for the gas flow layer [1]:

$$\alpha \propto (T_f - T_s)^{\frac{1}{4}} \qquad 5.7$$

Because of the low exponent the thermal transition coefficient is constant in first approximation over a large temperature range. Comparative calculations of the heating process lead to the conclusion that both approximations produced similar errors when fitted to an experimental temperature curve. Hence in the following the first approximation with constant α will be used.

5.2.3 Combined heating process

In the preceding sections the heat flux into the surface of the sample by radiation and convective heating has been discussed separately. Now both heat transport mechanisms shall contribute to the total heat flux into the sample:

$$\dot{Q} = \dot{Q}_{\text{radiation}} + \dot{Q}_{\text{conductive}} = \varepsilon_{\text{fs}} A_{\text{f}} \sigma (T_{\text{f}}^4 - T_{\text{s}}^4) + A_{\text{s}} \alpha (T_{\text{f}} - T_{s}) \quad 5.8$$

For the simulation a heat flux density is needed. Equation (5.8) is divided by A_{s} and a common thermal transition coefficient introduced:

$$\dot{q} = \frac{\dot{Q}}{A_{\text{s}}} = \alpha(T_{\text{f}}, T_{s})(T_{\text{f}} - T_{s}) \quad 5.9$$

with

$$\alpha(T_{\text{f}}, T_{\text{s}}) = \varepsilon_{\text{fs}} \frac{A_{\text{f}}}{A_{\text{s}}} \sigma (T_{\text{f}}^2 + T_{\text{s}}^2)(T_{\text{f}} + T_{\text{s}}) + \alpha \quad 5.10$$

Using the values introduced above, $\alpha(T_{\text{f}}, T_{\text{s}})$ is plotted in Fig. 5.1. Obviously at furnace temperatures above 900 °C heating by radiation is more important than the convective heat flux to the sample. It should also be noted that the

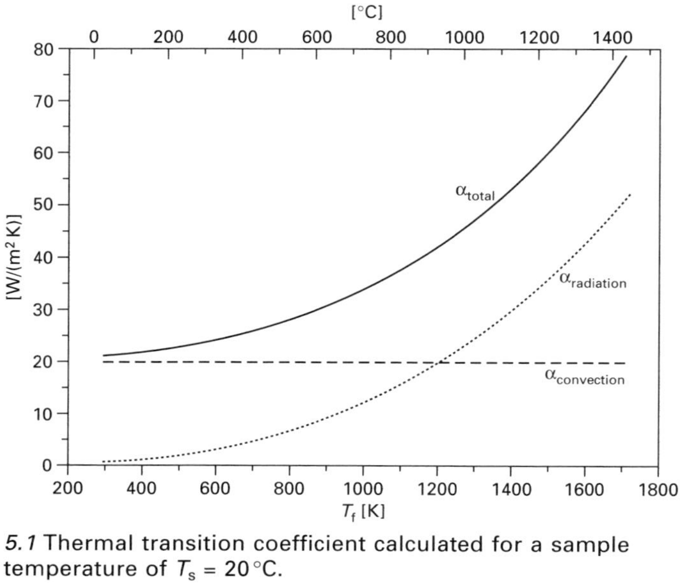

5.1 Thermal transition coefficient calculated for a sample temperature of T_{s} = 20 °C.

thermal transition coefficient doubles its value two times in the plotted temperature range.

Because the thermal conductivity of metallic samples is quite high, the temperature difference between the surface and the centre of the sample can be neglected in most cases Ceramic materials and thermal barrier coatings are consequently not covered by the following considerations. The heating process will now be calculated using a uniform sample temperature. The most important property of the sample is its heat capacity:

$$c(T) = \frac{\mathrm{d}Q}{\mathrm{d}T}\frac{1}{\rho V} \tag{5.11}$$

The heat transport from the furnace into the sample was defined by

$$\dot{Q} = A_s\alpha(T_f, T_s)(T_f - T_s) \tag{5.12}$$

Combining both equations, we get a simple integral rule to calculate the heating of the sample:

$$\mathrm{d}Q = A_s\alpha(T_f, T_s)(T_f - T_s)\mathrm{d}t = c(T_s)\rho V \mathrm{d}T_s \tag{5.13}$$

$$\frac{\mathrm{d}T_s}{\mathrm{d}t} = \frac{A_s\alpha(T_f, T_s)(T_f - T_s)}{c(T_s)\rho V} \tag{5.14}$$

$$T_s(t) = T_s(0) + \int_0^t \frac{A_s\alpha(T_f, T_s)(T_f - T_s)}{c(T_s)\rho V}\mathrm{d}t \tag{5.15}$$

The differential equation (5.14) cannot be solved in an analytical way, because $\alpha(T_f, T_s)$ introduces three terms with different orders in T_s. For each of the terms alone an analytical solution can be given, but the combination of the three makes this treatment impossible. Therefore the problem needs to be solved by numerical iteration, using equation (5.15). A time step size of 3.09 $\times 10^{-7}$ s was chosen. If the sample is treated as sheet material with thickness d the volume can be approximated, which leads to a surface to volume ratio (s) of half the material thickness 2.

Figure 5.2 was calculated using the values given in Table 5.1. The experimental heating curves were measured by M. Goebel in our laboratory and also published in Göbel *et al.* [5]. A thermocouple was placed inside a hole in a 2 mm thick sample to measure the sample core temperature. The calculated temperatures do not fit the measured curves exactly, but in relation to the large temperature range the correlation is better than using only radiative or convective heating for calculation. The calculated curves give at least a good impression of the heating times. In Göbel *et al.* [5] the faster heating to higher temperatures was related to special features of the furnace temperature control. Now it is obvious that the higher radiative heat transport at higher furnace temperatures is responsible for this effect.

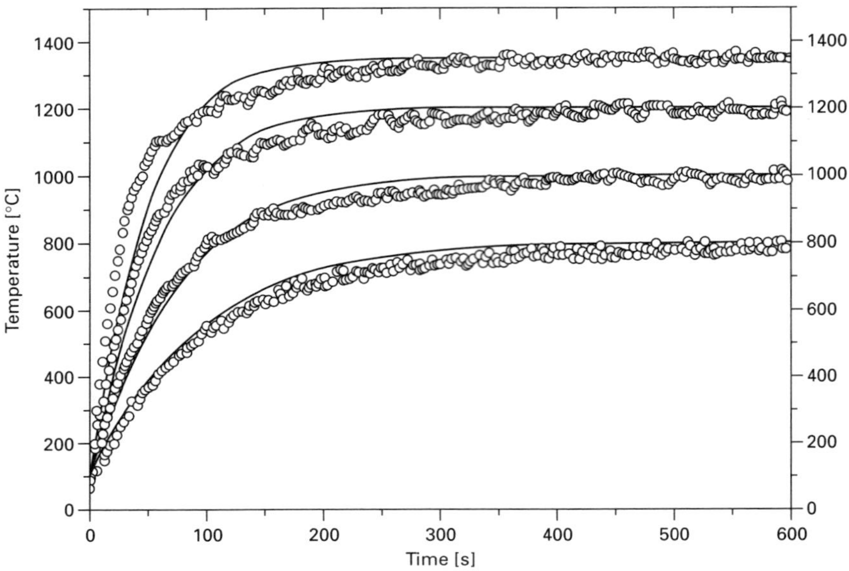

5.2 Calculated and measured heating curves for Fe20Cr5Al material.

Table 5.1 Values used for the calculated curves in Fig. 5.2

s	e_{fs}	α	A_f/A_s
1 mm	0.011	20 W/(m^{2}K)	14

5.3 Oxide growth under non-isothermal conditions

Following Kofstad [6] the growth law for oxides can generally be written in the form

$$x^n = \frac{dx}{dt} = \frac{k}{n+1} \qquad 5.16$$

where x is the oxide thickness, n an exponent describing the type of growth law (e.g. $n = 1$ for parabolic and $n = 0$ for linear growth) and k the growth constant, which is temperature dependent. It should be noted that n is usually a constant within certain temperature intervals,

$$k(T) = k_0 e^{-E_A/RT} \qquad 5.17$$

To obtain the kinetics of the oxide growth under non-isothermal conditions both equations need to be combined and integrated:

$$\int_{t=t_1}^{t=t_2} (n+1)x^n \frac{dx}{dt} dt = \int_{t=t_1}^{t=t_2} k_0 e^{-E_A/RT} dt \qquad 5.18$$

The left hand side integral can be solved easily:

$$\int_{t=t_1}^{t=t_2} (n+1)x^n \frac{dx}{dt} dt = \int_{x(t_1)}^{x(t_2)} (n+1)x^n dx$$

$$= x^{n+1}\Big|_{x(t_1)}^{x(t_2)} = x(t_2)^{n+1} - x(t_1)^{n+1} \qquad 5.19$$

The right hand side integral is more difficult, because the time dependence of the temperature is not yet defined. In some experiments the temperature is changed for purpose. Here constant heating rates are often used (e.g. refs [6]–[10]). For constant heating rates the model developed in Markworth [11] is very efficient and gives reliable results (see, e.g., [9]). In Liu and Gao [12] a numerical model is described, which divides the heating process into small time intervals and assumes parabolic growth within these intervals for a medium temperature. Thus non-linear temperature curves can also be treated. Later in this work the time dependence of the temperature determined in Section 5.2.3 will be used and the integral of the oxide growth (eq. (5.18)) will be treated by direct numerical integration.

5.3.1 Experimental verification on MA956

To validate the numerical solution of equation (5.18) the high-temperature alloy MA956 was chosen, to compare the results obtained by numerical calculation to non-isothermal experiments in the thermobalance. In the experiment the temperature was increased with 0.1 and 0.5 °C/min from 250 to 1300 °C (Fig. 5.3).

In Lesage *et al.* [13] the activation energy for the parabolic growth of alumina on MA956 is 266 kJ/mol with $k_0 = 1.2025 \times 10^{-2} g^2/(cm^4\ s)$. The experimental data could better be approximated by data obtained from isothermal oxidation experiments on PM 2000 – a similar ODS alloy – in the framework of the ALUSI* project, because the temperature range used in [13] covers only 80 °C. For PM 2000 the values are $k_0 = 8.5092 \times 10^{-1} g^2$ $(cm^4\ s)$ and $E_A = 320$ kJ/mol In our calculation the growth constant has slightly been adapted ($k_0 = 1.1210\, g^2/(cm^4\ s)$).

To obtain similar curves from experimental data the buoyancy effect in the hot air has first to be subtracted. The balance measures a mass m_m, which results from all forces acting on the supporting wire:

$$m_m = \frac{1}{g} \sum_i F_i \qquad 5.20$$

* Development of Alumina forming ODS ferritic superalloys as new biomaterials for Surgical Implants

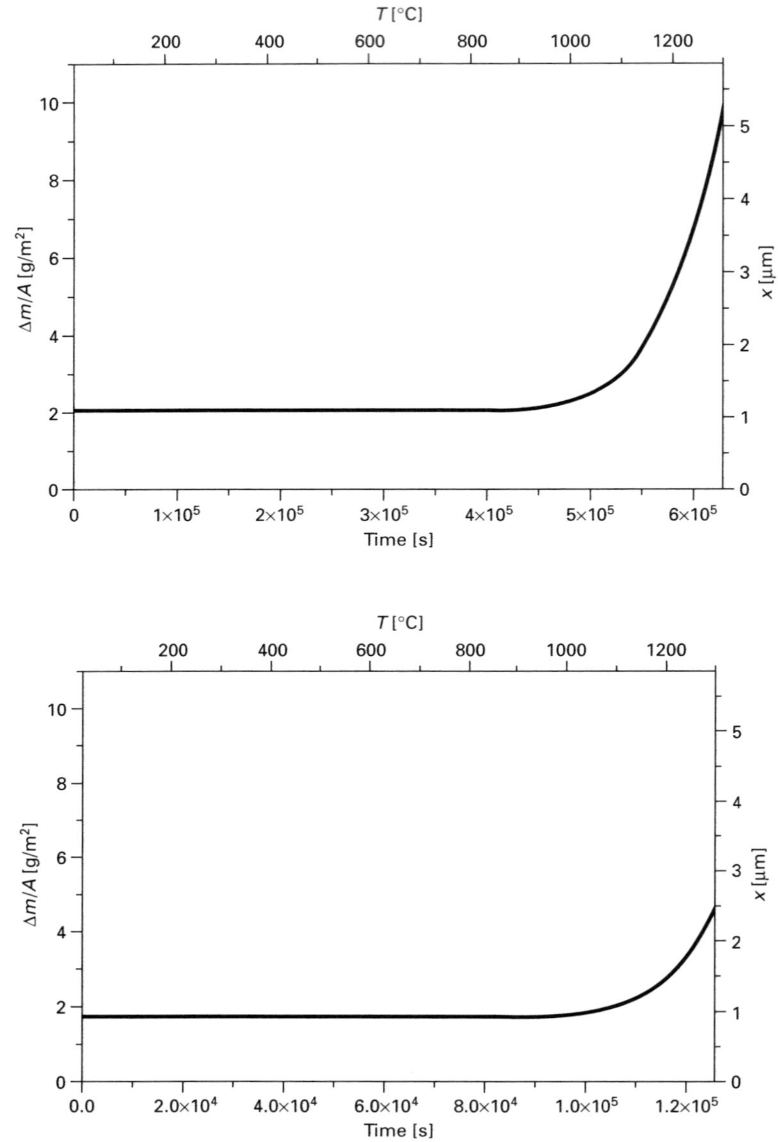

5.3 Oxide growth on MA956 calculated for constant heating rates of 0.1 and 0.5 K/min.

The forces comprise mainly the real weights of the sample and the holder as well as the buoyancy effect of both:

$$\sum_i F_i = gV_s\rho_s + gV_h\rho_h - gV_s\rho_a - gV_h\rho_a \tag{5.21}$$

V_s and V_h are the volumes cf the sample and the sample holder, respectively. ρ_s, ρ_h, ρ_a are the corresponding densities and that of air. Rearranging the sum a buoyancy mass (m_b), the sample mass (m_s) and the mass of the holder (m_h) can be defined:

$$m_s V_s \rho_s \tag{5.22}$$

$$m_h = V_h \rho_h \tag{5.23}$$

$$m_b = V_s + V_h)\rho_a \tag{5.24}$$

$$m_m = m_s + m_h - m_b \tag{5.25}$$

To calculate the density of air the perfect gas equation

$$pV = nRT \tag{5.26}$$

and a standard composition of air

$$m_a = m_{O_2} + m_{N_2} = 0.2n2M_O + 0.8n2M_N = nM_a = n28.81\,\text{g/mol} \tag{5.27}$$

is used:

$$\rho_a = \frac{m_a}{V} = \frac{nM_a}{V} = M_a \frac{n}{V} = M_a \frac{p}{RT} \tag{5.28}$$

Now the buoyancy mass is:

$$m_b = (V_s + V_h)\frac{pM_a}{RT} \tag{5.29}$$

The nominal sample dimensions are 20 mm, 10 mm and 2 mm. The volume is $V_s = 4 \times 10^{-7}\,\text{m}^3$. (For the evaluation of the measurement the exact volumes from Table 5.2 were usec.) For $V_h = 2.0 \times 10^{-6}\,\text{m}^3$ was determined. The buoyancy corrected mass gain from the experiments is shown in Fig. 5.4. Note that in Clausthal-Zellerfeld, which is at an altitude of 535–610 m, the pressure has to be reduced by 65 mbar (at 570 m) to 948 mbar. Taking all this together gives 0.7884 g K mol.

To ensure that during the non-isothermal oxidation only alumina is formed and thus the growth law given above is applicable, the samples were preoxidised for 4 h at 1100 °C. The mass gains after the preoxidation and after the non-isothermal experiment are given in Table 5.2.

A comparison of the measurement (Fig. 5.4) and the calculation (Fig. 5.3) shows that both are in good agreement. The onset of oxidation at approx. 1000 °C shows that the activation energy is correct. The good agreement in the overall mass gain indicates that the growth constant was chosen correctly. In the experiment with the heating rate of 0.1 K/min the overall mass gain was calculated a little bit higher than in the experiment. For the experiment

Table 5.2 Basic data of the samples

Sample ID	Length (mm)	Width (mm)	Thickness (mm)	Initial mass (g)	After pre-oxidation (g)	After experiment (g)	A (m^2)	V_s (m^3)	m/A (g/m^2)	m/A (g/m^2)
M0083P3	20.0	9.90	1.95	2.6029	2.6038	2.6058	5.1261×10^{-4}	3.861E-7	1.755	5.657
M0016P4	20.1	10.15	1.95	2.6789	2.6800	2.6839	5.2601×10^{-4}	3.978E-7	2.091	9.506

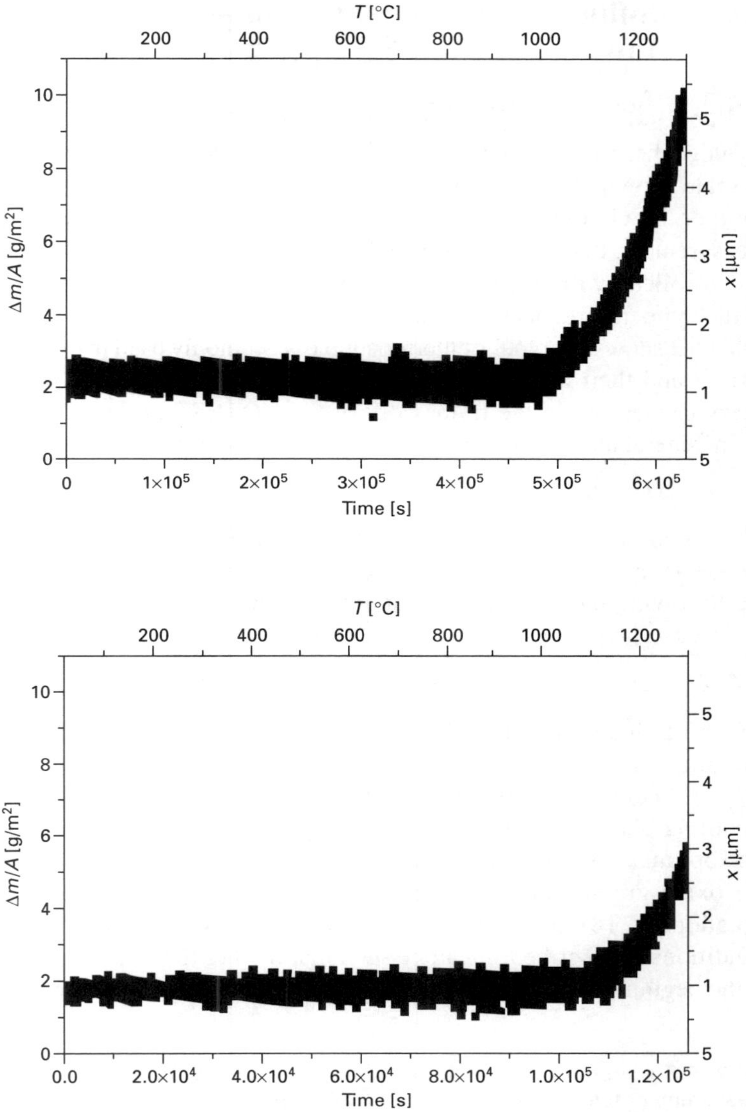

5.4 Oxide growth on MA956 measured for constant heating rates of 0.1 and 0.5 K/min. The buoyancy effect has been subtracted.

with the heating rate of 0.5 K/min the relation is inverted. Obviously the growth law is not exactly parabolic. Nevertheless the good correspondence between both figures justifies the use of a numerical calculation method for an estimation of the oxide growth during the heating phase in cyclic oxidation experiments.

5.4 Influence of the heating phase on the oxidation time

5.4.1 Accuracy of the temperature measurement

To judge the error induced by oxidation times which may or may not contribute to scale growth during the heating period it is necessary to compare it with another inevitable error induced by the inaccuracy of the temperature measurement. Because in nearly all high-temperature furnaces the temperature is controlled by thermocouples, it is the most direct way to use the accuracy of the temperature measurement as laid down in international standards [14]. Table 5.3 shows the types of thermocouples commonly used in cyclic oxidation testing and their accuracies. The selection of thermocouples is based on a query carried out in the framework of the COTEST project.

In general the error of a temperature measurement can be calculated using:

$$\Delta T = aT \quad 5.30$$

where a are the linear tolerance factors as given in Table 5.3 depending on the thermocouple used. Note that in the standards T is given in °C, while in the following the Kelvin temperature scale will be used.

5.4.2 Estimation of allowable heating time

For the estimation of the effect of heating on the total oxidation time the oxidation during the defined cycle time (t_{cycle}) with a constant temperature (T_{cycle}) is compared with the oxidation during a real cycle with a heating and a cooling phase. At first only the heating phase is taken into consideration. The optimal experimental cycle time with heating phase would be such that the oxide grown during the theoretical and the real cycle is the same. In equations (5.18) and (5.19) the growth of an oxide under non-isothermal conditions is described. It can be seen that neither the initial oxide thickness at the beginning of the cycle nor the exponent of the growth law is important,

Table 5.3 Commonly used thermocouples in cyclic oxidation and maximum measurement tolerances following DIN IEC 60584,2

Thermo-couple type	Pair	Temperature range (°C)	Accuracy: Fixed value	Accuracy: Related to reading *t* (°C)
S	Pt-10%Rh/Pt	0–1600	±1.5 °C	±0.0025*t*
R	Pt-13%Rh/Pt	0–1600	±1.5 °C	±0.0025*t*
K	Ni-Cr/Ni-Al	−40–1200	±2.5 °C	±0.0075*t*
B	Pt-30%Rh/Pt-6%Rh	600–1700	±4.0 °C	±0.0050*t*

but only the value of the integral on the right hand side of (5.18). Thus the problem to be solved is to define in such a way that:

$$\int_0^{t_{\text{cycle}}^{\text{heat}}} e^{-E_A/RT(t)}\,\mathrm{d}t = \int_0^{t_{\text{cycle}}} e^{-E_A/RT_{\text{cycle}}}\,\mathrm{d}t \qquad 8.31$$

Because during the real cycle oxide growth occurs already during the heating phase at temperatures below T_{cycle} the real cycle time will always be longer than the defined cycle time. This additional time shall be called:

$$t_{\text{cycle}}^{\text{heat}} = t_{\text{cycle}} + t_{\text{cycle}}^{\text{add}} \qquad 5.32$$

To achieve standardisation it would even be better to give the relative temperature decrease (r) from the desired cycle temperature, to have a uniform criterion, independent of the actual cycle temperature. Hence r is defined as:

$$r = \frac{T_{\text{cycle}} - T(t_{\text{cycle}}^{\text{add}})}{T_{\text{cycle}}} \qquad 5.33$$

Some values can now be calculated. The sample dimensions are chosen close to the recommendations in Saunders and Nicholls [2]. In Table 5.4 the first three lines show recommended values, followed by two other often used methods of sample cutting.

For the calculation three values of the volume to surface ratio, covering the whole range, are chosen: 1.6 mm, 1.2 mm and 0.4 mm. All other parameters are chosen as in Section 5.2.3, starting the heating at 20 °C. The cycle times are 20 h for the long dwell time and 1 h for the short dwell time. The ultra-short dwell time (5 min) is not taken into consideration here, because heating method and sample dimensions are still a point of discussion. To save calculation time in the following only the short dwell time (1 h) was taken into consideration, because it is obvious that the additional cycle time is independent of the cycle time, if the latter is much larger than the heating time. Using the sample dimensions of the recommended block (Table 5.4, row 3), activation energies for the oxide growth of MA956 as a typical

Table 5.4 Sample dimensions for cyclic oxidation

Type	*d, w* (mm)	*l, h* (mm)	*t* (mm)	*V* (mm^3)	*A* (mm^2)	*V/A* (mm)
Rod	8	15	–	753.98	477.52	1.579
Disc	14	4	–	615.75	483.81	1.273
Block	15	10	4	600	500	1.2
Disc	17	1	–	226.98	507.37	0.447
Block	20	10	1	200	460	0.435

alumina former ($E_A = 320$ kJ/mol, see Section 5.3.1), activation energies for chromia formers ($E_A = 223$ kJ/mol, [15]), and for comparison purposes an even lower activation energy of 150 kJ/mol, values can be given.

To obtain the corrected cycle times the integral (5.31) was solved numerically. The temperatures were taken from the heating curves. An example is given in Fig. 5.5. Though the heating times are rather different, the temperatures obtained for different volume to surface ratios at the same temperature and for the same activation energy are similar. Figure 5.6 gives an overall picture, showing the relative temperature decrease (r) for three activation energies and the three chosen volume to surface ratios. It can also be seen that the influence of the heating phase is more important than the accuracy of the employed thermocouple.

Figure 5.7 shows that the data can be further simplified. The fraction of the Boltzmann factors at the cycle temperature gives a constant:

$$f = \frac{e^{-E_A/RT(t_{\mathrm{cycle}}^{\mathrm{add}})}}{e^{-E_A/RT_{\mathrm{cycle}}}} \approx 0.574 = \text{constant} \qquad 5.34$$

The decreased temperature can now be calculated easily in relation to the cycle temperature and the activation energy. The characteristics of the heating process are summarised in the constant f and those of the oxidation process in the activation energy E_A:

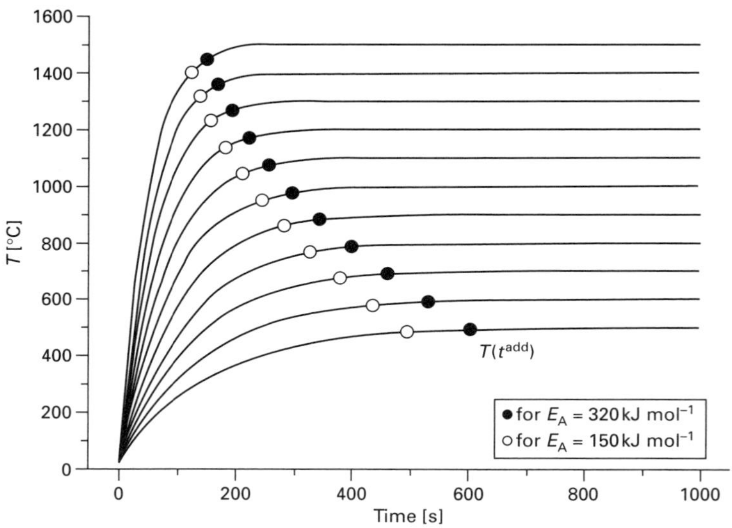

5.5 Heating curves for a volume to surface ratio of 1.2 mm, showing also the temperatures at $t_{\mathrm{cycle}}^{\mathrm{add}}$ calculated for two different activation energies.

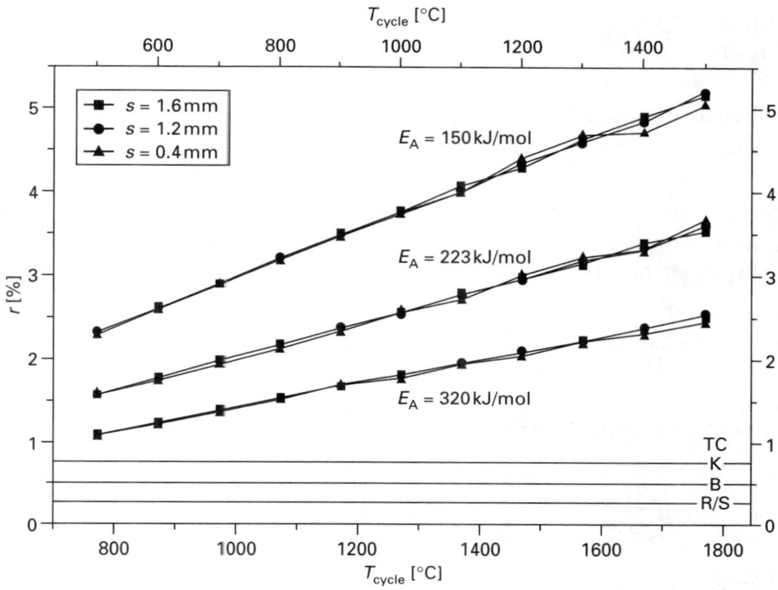

5.6 Relative temperature decrease for different activation energies and cycle temperatures.

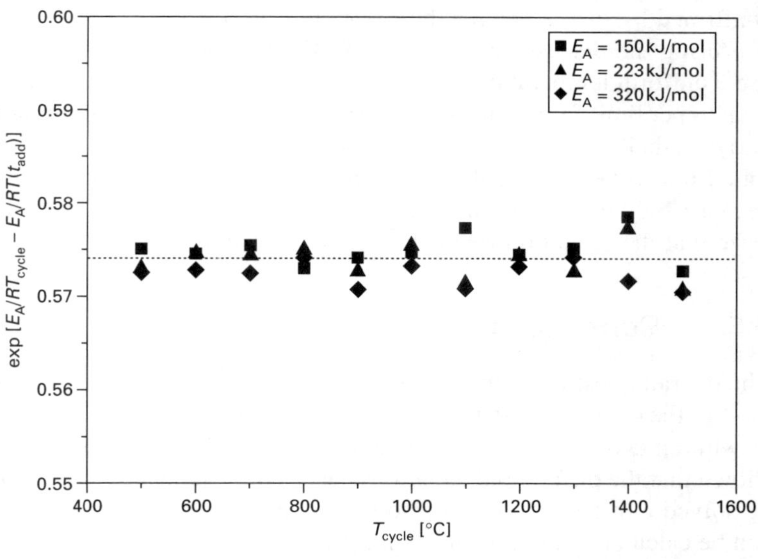

5.7 Fraction of the Boltzmann factor at the furnace temperature and the temperature at t_{cycle}^{add}.

$$T(t_{\mathrm{cycle}}^{\mathrm{add}}) = \frac{E_{\mathrm{A}}}{E_{\mathrm{A}} - RT_{\mathrm{cycle}} \ln f} T_{\mathrm{cycle}} \qquad 5.35$$

The same is true for the relative temperature decrease:

$$r = \frac{-RT_{\mathrm{cycle}} \ln f}{E_{\mathrm{A}} - RT_{\mathrm{cycle}} \ln f} \qquad 5.36$$

The slope of r is given by:

$$\frac{\partial r}{\partial T_{\mathrm{cycle}}} = \frac{-RE_{\mathrm{A}} \ln f}{(E_{\mathrm{A}} - RT_{\mathrm{cycle}} \ln f)^2} \qquad 5.37$$

and can be approximated by:

$$\frac{\partial r}{\partial T_{\mathrm{cycle}}} \approx \frac{-R \ln f}{E_{\mathrm{A}}} \approx \frac{R}{2E_{\mathrm{A}}} \qquad 5.38$$

using:

$$RT_{\mathrm{cycle}} << E_{\mathrm{A}} \quad \text{and} \quad \ln f = -0.5513 \approx -\frac{1}{2} \qquad 5.39$$

as it is the case for chromia- and alumina-forming alloys. This result is confirmed by the values for the slopes in Fig. 5.6.

During the discussion within the COTEST consortium it was agreed to use a uniform temperature decrease of $r = 3\%$ for all materials, because this is an upper limit for the most often employed chromia- and alumina-forming alloys in their respective service temperature ranges. At the same time it was found not to be worth defining a similar criterion for the cooling process, because the temperature drop at the beginning of the cooling phase is so large that the oxidation seems to stop immediately.

5.5 Conclusion

The overall question of this chapter was whether there is a simple method to modify the cycle time, so that the same oxide growth could be observed with or without taking the heating process of the sample into account. It has been shown that the differential equations describing the heating process can only be solved numerically. It has also been demonstrated that the oxide growth can be calculated numerically as long as the heating curve is known. On the basis of this knowledge, corrected cycle times have been calculated for the alumina-former MA956 with an activation energy for the parabolic growth law of 320 kJ/mol. For chromia formers the activation energies are in the same order of magnitude (223 kJ/mol [15]). As chromia formers are used up to 1000 °C and alumina formers in the temperature range above an upper limit of the relative temperature decrease can be defined for both classes of materials at 3%.

Within the COTEST consortium it was agreed to define the starting point of the cooling phase as the point of time when the sample leaves the furnace. The cooling phase will end when the sample temperature reaches 50 °C. This is close to ambient temperature and can be reached also under the conditions of a very hot climate. At ambient temperature all effects of the thermal mismatch between the protective layer and the substrate are expected to reach an equilibrium state. An example of such an oxidation cycle and the defined time periods is given in Fig. 5.8. In practice the easiest way to determine the duration of the four phases of an oxidation cycle is to measure the temperature curve with a thermocouple directly attached to a dummy sample.

5.6 Acknowledgements

The data on oxidation kinetics for PM 2000 were collected in the framework of the ALUSI project (Development of alumina forming ODS ferritic superalloys as new biomaterials for surgical implants) funded under EC contract No. G5RD-CT1999-00083. The idea for the topic of this publication

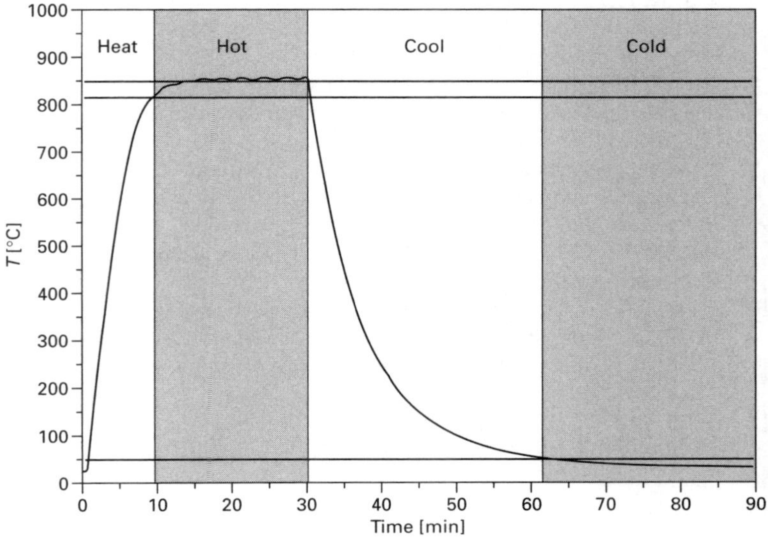

5.8 Example of an oxidation cycle with typical temperature profile and the four phases of heating, hot dwell time, cooling and cold dwell time.

and some of the data as marked in the text are based on the COTEST project (Cyclic oxidation testing – development of a code of practice for the characterisation of high temperature materials performance) funded under EC contract No. G6RD-CT-2001-00639. Also special thanks to Dr M. Malessa from DECHEMA, who performed the measurement of the temperature cycle shown in Fig. 5.8.

5.7 References

1. H. D. Baehr, and K. Stephan, *Wärme- und Stoffübertragung*, Springer-Verlag, Berlin, Heidelberg, New York, 1998.
2. S. R. J. Saunders, and J. R. Nicholls, Code of practice for discontinuous corrosion testing in high temperature gaseous atmospheres, Consortium of the EC-Project 'TESTCORR', CEC contract no. SMT4-CT95-2001, 2000.
3. E. Hering, R. Martin and M. Stohrer, *Physik für Ingenieure*, Springer Verlag, Berlin, Heidelberg, New York, 2002.
4. U. Grigull, H. Gröber, and S. Erk, *Die Grundgesetze der Wärmeübertragung*, Springer-Verlag, Berlin, Göttingen, Heidelberg, 1963, 3. Auflage
5. M. Göbel, A. Glaskov, M. Konopka, J. Jedlinski, G. Borchardt, and J. Le Coze, Influence of start-up oxidation procedure on the composition of oxide scales of high temperature alloys during initial oxidation, in: S. B. Newcomb and J. A. Little, editors, *Microscopy of Oxidation* 3, pages 12–18, The Institute of Materials, 1997, held at Trinity Hall, The University of Cambridge, 16–18 September 1996.
6. P. Kofstad, Oxidation of metals: Determination of activation energies, *Acta Chemica Scandinavica*, **12**(4): 701–707, 1958.
7. M. U. Kitheri, P. Srirama Murti, and G. Seenivasan, Thermoanalytical study of the oxidation of selected metals with reference to the influence of heating rate, *Thermochimica Acta*, **232**(1): 129–136, 1994.
8. A. N. Dil'din, B. P. Belozerov, G. G. Mikhailov, and B. I. Leonovich, Kinetics of nonisothermal oxidation of alloy steel components, *Steel in Translation*, **24**(6): 15–17, 1994.
9. R. K. Singh Raman, B. Gleeson, and D. J. Young, Theoretical prediction and experimental measurement of anisothermal oxidation kinetics, in: *Proceedings of the 13th International Corrosion Congress: Vol. III*, pages 297/1–297/7, Australasian Corrosion Association Inc., 1996, held 25–29 November 1996, in Melbourne, Australia
10. Z. Liu, W. Gao, and H. Gong, Anisothermal oxidation of micro-crystalline Ni-20Cr-5Al alloy coating at 850–1280 °C, *Scripta Materialia*, **38**(7): 1057–1063, 1998.
11. A. J. Markworth, On the kinetics of anisothermal oxidation, *Metallurgical Transactions A*, **8A**: 2014–2015, 1977.
12. Z. Liu and W. Gao, A numerical model to predict the kinetics of anisothermal oxidation of metals, *High Temperature Materials and Processes*, **17**(4): 231–236, 1998.
13. B. Lesage, L. Maréchal, A. M. Huntz, and R. Molins, Aluminium depletion in FeCrAl alloys during oxidation, *Defect and Diffusion Forum*, **194–199**: 1707–1712, 2001.

14. DIN EN 60584 und 43722, Thermopaare, Teil 1-3, Beuth Verlag, IEC 60584,1-3 modified.
15. S. Chevalier, G. Bonnet, K. Przybylski, J. C. Colson, and J. P. Larpin, Segregation of neodymium in chromia grain-boundaries during high-temperature oxidation of neodymium oxide-coated chromia-forming alloys, *Oxidation of Metals*, **54**(5): 527–548, 2000.

6

Investigation of the influence of parameter variation in long dwell thermal cycling oxidation

L. NIEWOLAK and W. J. QUADAKKERS,
Forschungzentrum Jülich, Germany

Long dwell time tests find wide application in industry, a major market being power generation and energy conversion industries (steam and gas turbines, power plants, fuel cells, etc.). Parts and structures operating under certain conditions are used across a wide sector of the manufacturing industry, including alloy manufacture, power plants, petrol and chemical industry, commercial heat treatment, coating manufacture, pottery, automotive and aerospace industries. Other applications are also of importance, particularly foil-based automotive catalyst carriers, burners, hot gas filters, airplane engines, etc. These applications play an important role in the power generation, chemical, automotive and aerospace industries. Such components are manufactured as cast, forged or rolled products, but not necessarily so. Thus a major objective of the 'long dwell test standardisation' was to define a test method with wide applicability to all industrial sectors.

6.1 Introduction

This contribution describes all experimental data obtained by Forschungszentrum Jülich in the frame of the COTEST project. Table 6.1 shows the test matrix of experiments that have been performed to investigate the influence of parameter variation on the oxidation behaviour in long dwell thermal cycling oxidation.

6.2 Experimental set-up

The following materials were used in the test programme: P91, AISI 441, Alloy 800H, CM 247 and Kanthal A1. For the oxidation tests, samples of the selected materials with nominal dimensions $12 \times 17 \times 1.5\,mm^3$ were machined and subsequently ground down to 1200 grit surface finish. Additionally, half of the ground P91 specimens were heat-treated for 3 h at 1000 °C in an Ar–5%H_2 atmosphere. The aim of this procedure was to enhance the oxidation

Table 6.1 Test matrix

Test	Material	Upper dwell time/ lower dwell time (h/h)	Temperature (°C)	Gas
FZJ-1	P91	4/4	650	Dry
FZJ-2	P91	20/4	650	Wet
FZJ-3	P91	4/4	650	Wet
FZJ-4	P91	20/4	650	Dry
FZJ-5	AISI 441	4/2	800	Dry
FZJ-6	AISI 441	4/4	850	Dry
FZJ-7	AISI 441	4/4	900	Wet
FZJ-8	AISI 441	8/4	800	Wet
FZJ-9	AISI 441	8/4	850	Dry
FZJ-10	AISI 441	8/2	900	Dry
FZJ-11	AISI 441	20/4	800	Dry
FZJ-12	AISI 441	20/2	850	Wet
FZJ-13	AISI 441	20/4	900	Dry
FZJ-14	Alloy 800H	4/2	950	Dry
FZJ-15	Alloy 800H	4/4	1000	Dry
FZJ-16	Alloy 800H	4/4	1050	Wet
FZJ-17	Alloy 800H	8/4	950	Wet
FZJ-18	Alloy 800H	8/4	1000	Dry
FZJ-19	Alloy 800H	8/2	1050	Dry
FZJ-20	Alloy 800H	20/4	950	Dry
FZJ-21	Alloy 800H	20/2	1000	Wet
FZJ-22	Alloy 800H	20/4	1050	Dry
FZJ-23	CM 247	4/4	1100	Dry
FZJ-24	CM 247	20/4	1150	Dry
FZJ-25	CM 247	4/4	1150	Dry
FZJ-26	CM 247	20/4	1100	Dry
FZJ-27	Kanthal A1	4/4	1200	Dry
FZJ-28	Kanthal A1	20/4	1250	Dry
FZJ-29	Kanthal A1	4/4	1250	Dry
FZJ-30	Kanthal A1	20/4	1200	Dry

rate by reducing the chromium diffusion in the surface deformation zone. Figure 6.1 shows microstructures of the tested materials.

The oxidation experiments were carried out in dry synthetic air (N_2 + 20% vol. O_2 with less than 10 ppm of water vapour) or wet synthetic air with 2% specific humidity. The oxidation experiments were conducted in a specially designed horizontal furnace facility (Fig. 6.2).

6.3 Experimental results

6.3.1 Oxidation of steel P91

Figures 6.3–6.12 show net and gross weight changes for steel P91 during cyclic oxidation at 650 °C. During each experiment four specimens of steel

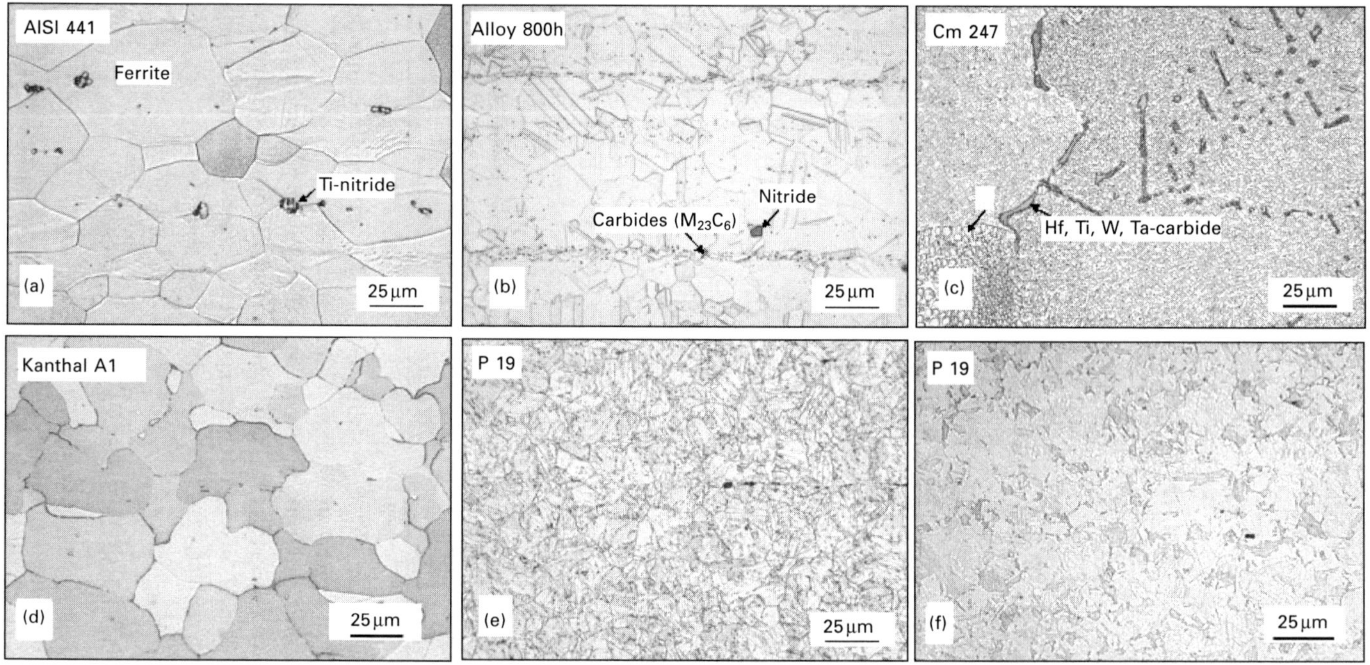

6.1 Optical metallography of the microstructures of the alloys investigated in the 'as received' state (a) AISI 441, (b) Alloy 800H, (c) CM 247, (d) Kanthal A1, (e) P91, (f) P91 after heat treatment for 3 h at 1000 °C in Ar–5%H_2 gas.

(a)

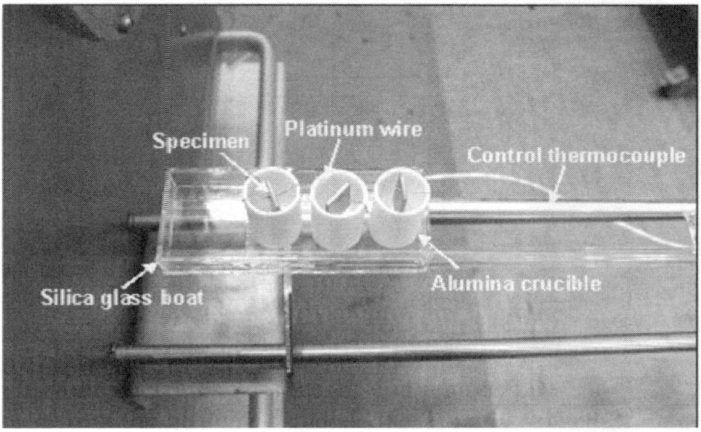

(a)

6.2 (a) Cyclic oxidation furnace, (b) specimen arrangement.

P91 were oxidised; two of them had been heat treated and the other two were in the 'as-received' state. The non-heat-treated P91 specimens (Figs 6.3–6.12) during oxidation in dry and wet air as well as the heat-treated specimens during oxidation in dry air showed almost no mass change. These negligible mass changes result from the formation of an extremely thin, protective oxide scale during the cyclic testing (Figs 6.13a and 6.14a). The X-ray diffraction (XRD) and sputtered neutral mass spectrometry (SNMS) measurements revealed that the oxide scale consisted of $(Fe, Cr)_2O_3$ and $(Cr, Mn)_3O_4$ spinel (Figs 6.15 and 6.16). In contrast, an enhanced oxidation as the result of a breakaway process was observed for the heat-treated material

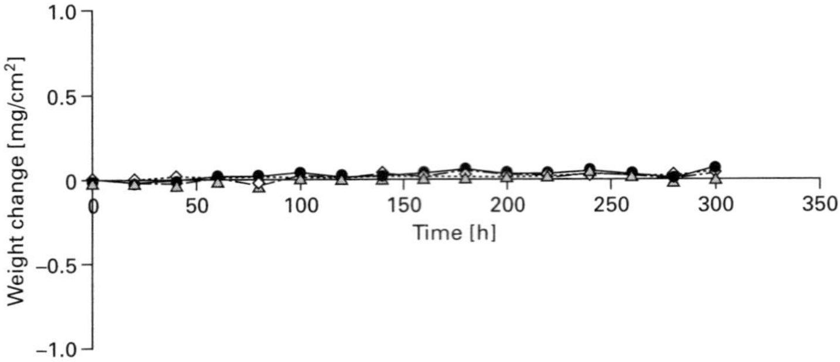

6.3 Net weight changes for steel P91 (non-heat treated) during cyclic oxidation (4 h/4 h cycles) for 300 h at 650 °C in dry air (FZJ-1).

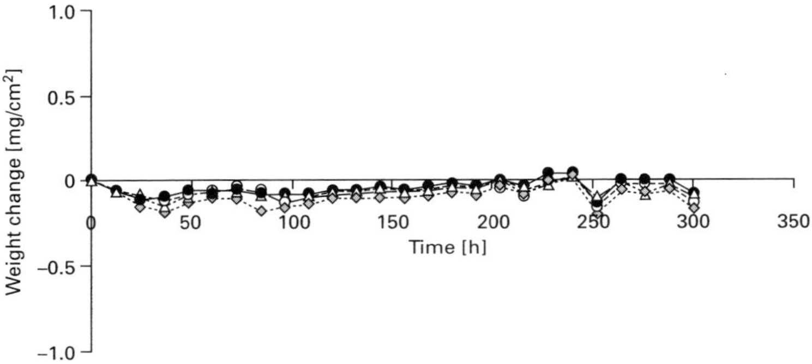

6.4 Gross weight changes for steel P91 (non-heat treated) during cyclic oxidation (4 h/4 h cycles) for 300 h at 650 °C in dry air (FZJ-1).

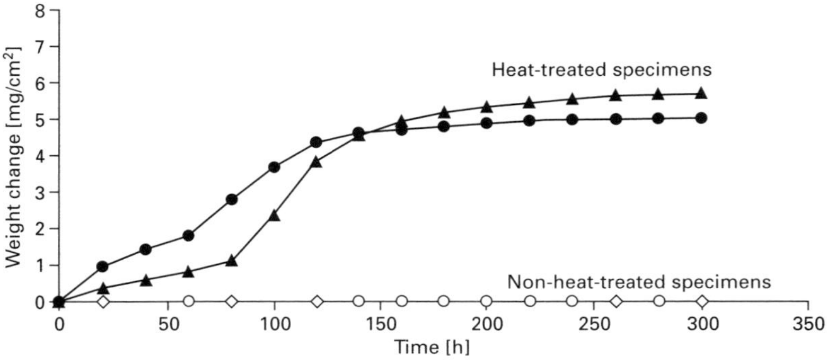

6.5 Net weight changes for steel P91 during cyclic oxidation (20 h/4 h cycles) for 300 h at 650 °C in wet air (FZJ-2).

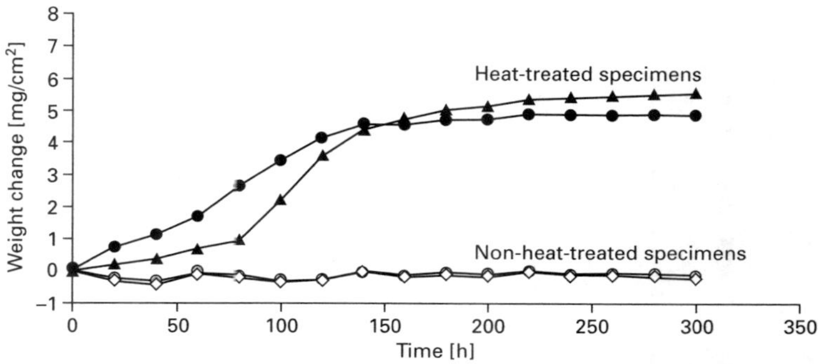

6.6 Gross weight changes for steel P91 during cyclic oxidation (20 h/4 h cycles) for 300 h at 650 °C in wet air.

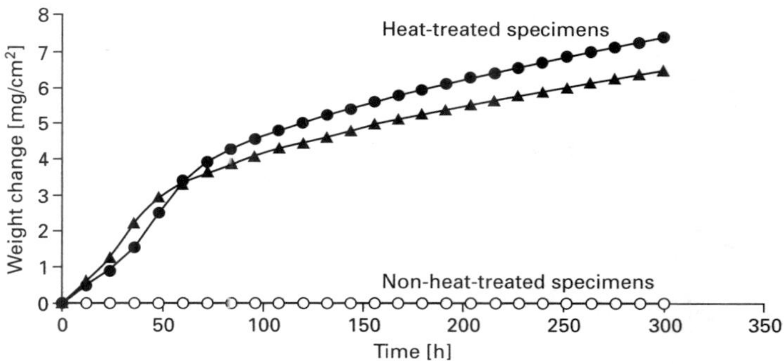

6.7 Net weight changes for steel P91 during cyclic oxidation (4 h/4 h cycles) for 300 h at 650 °C in wet air (FZJ-3).

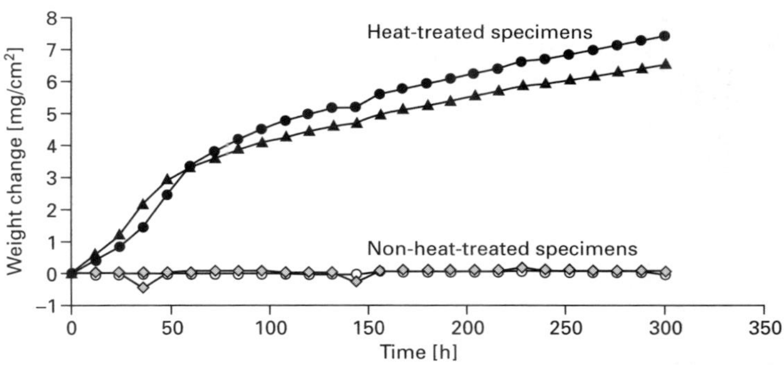

6.8 Gross weight changes for steel P91 during cyclic oxidation (4 h/4 h cycles) for 300 h at 650 °C in wet air (FZJ-3).

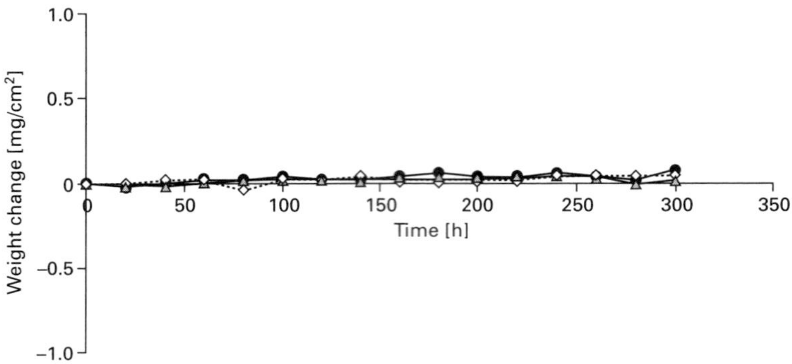

6.9 Net weight changes for steel P91 (non-heat-treated) during cyclic oxidation (20 h/4 h cycles) for 300 h at 650 °C in dry air (FZJ-4).

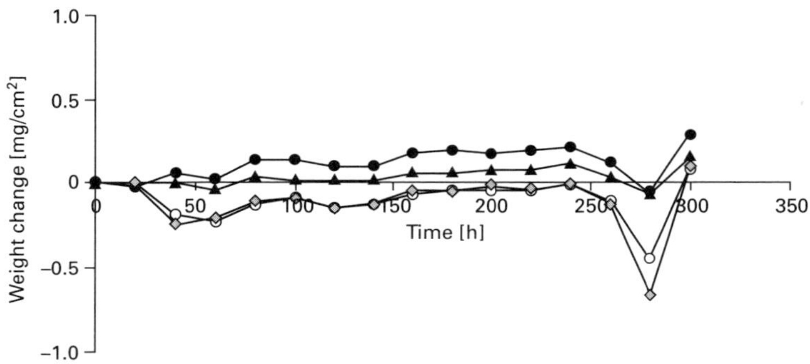

6.10 Gross weight changes for steel P91 (non-heat-treated) during cyclic oxidation (20 h/4 h cycles) for 300 h at 650 °C in dry air (FZJ-4).

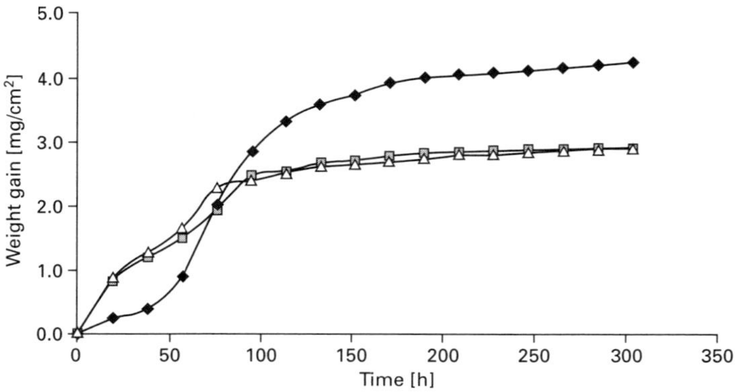

6.11 Net weight changes for steel P91 (heat-treated specimen) during cyclic oxidation (1 h/0.25 h cycles) for 300 h at 650 °C in wet air (FZJ-31).

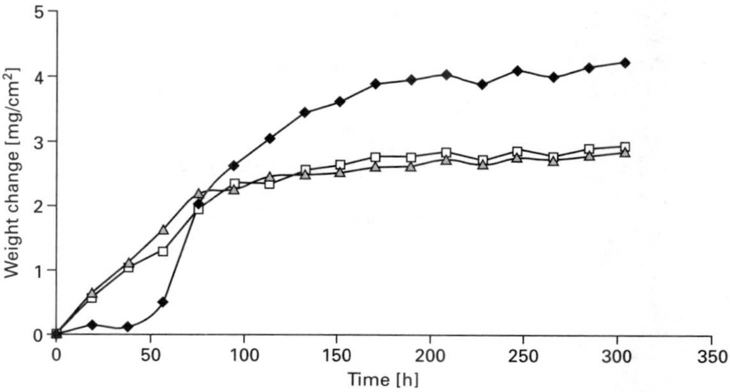

6.12 Gross weight changes for steel P91 (heat-treated specimen) during cyclic oxidation (1 h/0.25 h cycles) for 300 h at 650 °C in wet air (FZJ-31).

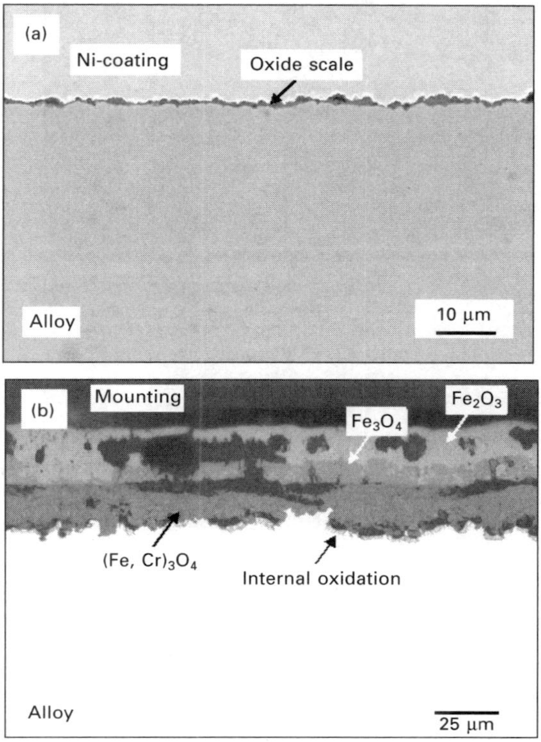

6.13 Microstructures of the oxide scales on P91 after 300 h cyclic oxidation at 650 °C. (a) scanning electron microscopy/backscatter electron (SEM/BSE) image of the non-heat-treated specimen – 20 h/4 h, dry air; (b) optical metallography image of the heat-treated specimen – 4 h/4 h, wet air.

6.14 SEM images of the oxide scale surfaces formed on steel P91 during 300 h cyclic oxidation (4 h/4 h) in wet air at 650 °C: (a) non-heat-treated specimen, (b) heat-treated specimen.

in wet air (Figs 6.13b and 6.14b). The non-protective scale consisted of an outer layer of haematite (with haematite whiskers (Fig. 6.14b), an inner layer of Fe_3O_4 (magnetite) and $(Fe, Cr)_3O_4$ spinel and a narrow internal oxidation zone (Fig. 6.13b).

6.3.2 Oxidation of steel AISI 441

The net and gross mass change curves for AISI 441 exposed at 800, 850 and 900 °C are given in Figs 6.17–6.34. Increasing the temperature resulted in an increase of the oxidation rate. In the early stages of exposure, the material's

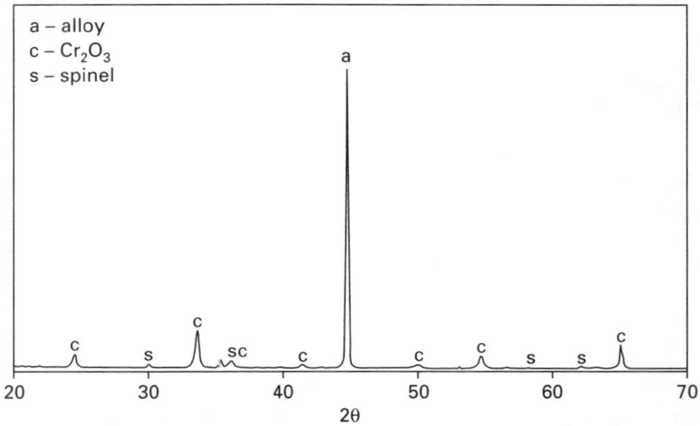

6.15 XRD pattern of oxide scale on steel P91 (heat-treated specimen) after cyclic oxidation (20 h/4 h cycles) for 300 h at 650 °C in dry air.

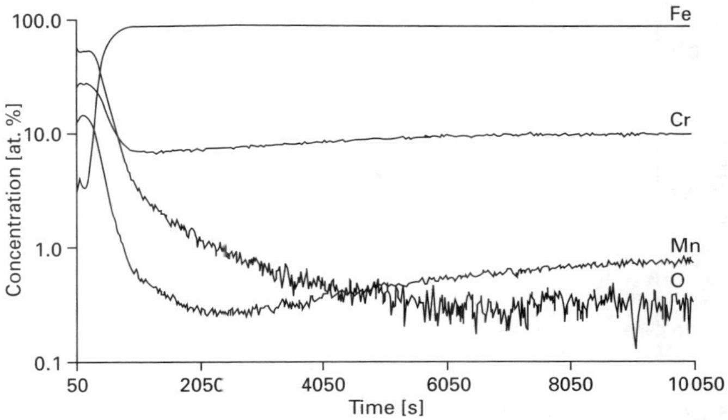

6.16 SNMS depth profile of non-heat-treated steel P91 after 300 h cyclic oxidation (20 h/4 h cycles) at 650 °C in dry air.

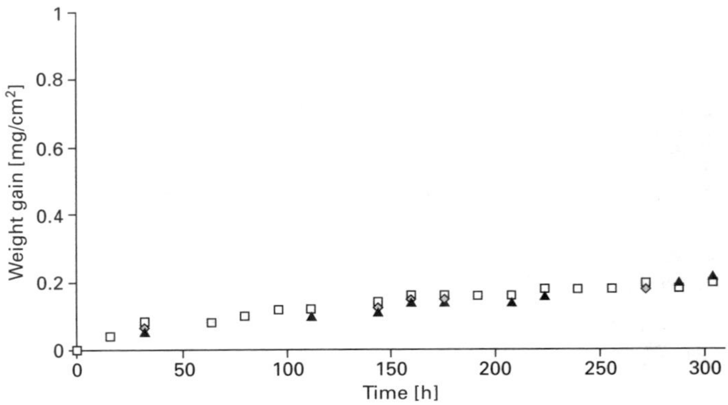

6.17 Net weight changes for steel AISI 441 during cyclic oxidation (4 h/2 h cycles) for 300 h at 800 °C in dry air (FZJ-5).

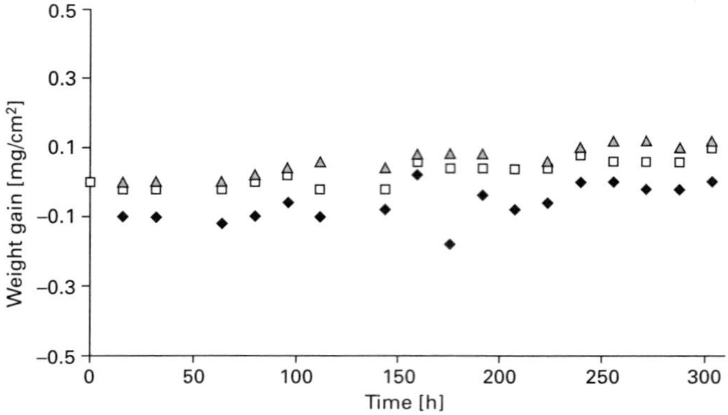

6.18 Gross weight changes for steel AISI 441 during cyclic oxidation (4h/2h cycles) for 300h at 800°C in dry air (FZJ-5).

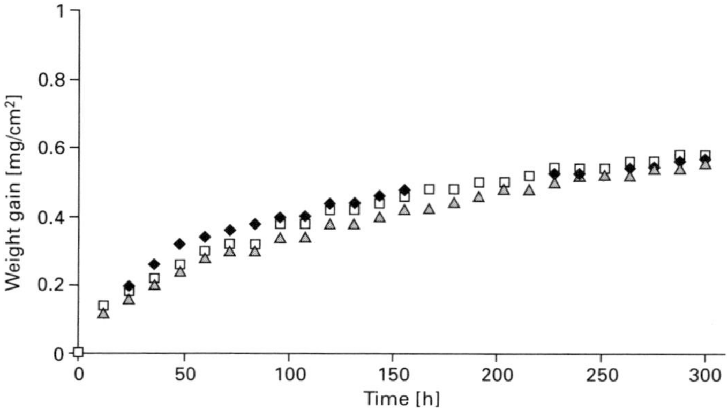

6.19 Net weight changes for steel AISI 441 during cyclic oxidation (4h/4h cycles) for 300h at 850°C in dry air (FZJ-6).

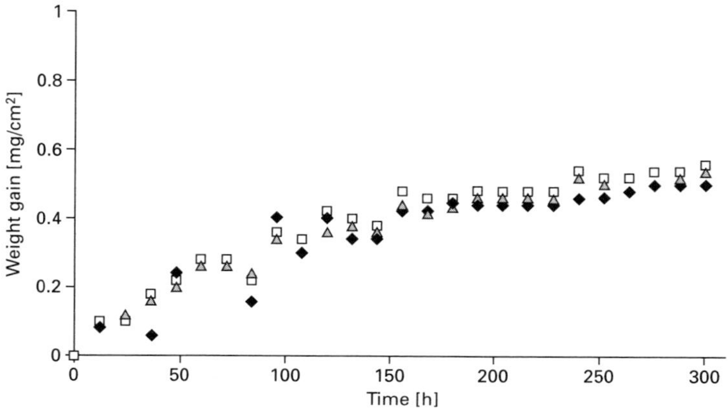

6.20 Gross weight changes for steel AISI 441 during cyclic oxidation (4h/4h cycles) for 300h at 850°C in dry air (FZJ-6).

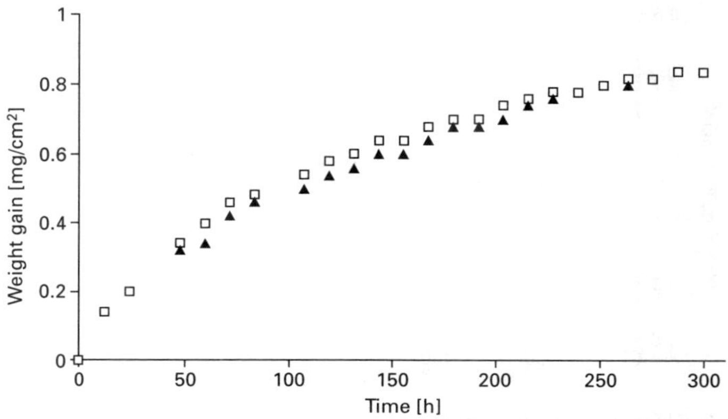

6.21 Net weight changes for steel AISI 441 during cyclic oxidation (4 h/4 h cycles) for 300 h at 900 °C in wet air (FZJ-7).

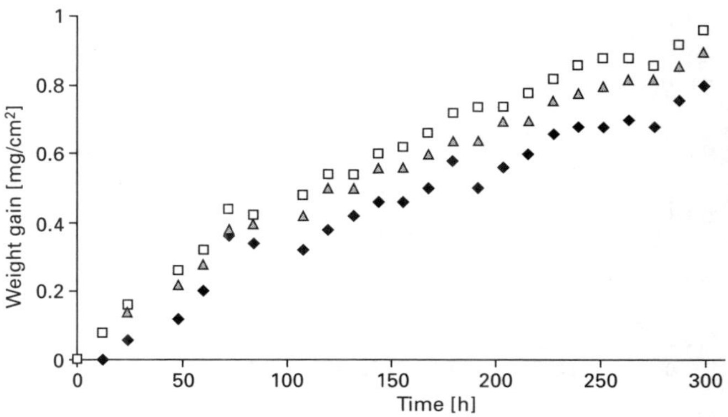

6.22 Gross weight changes for steel AISI 441 during cyclic oxidation (4 h/4 h cycles) for 300 h at 900 °C in wet air (FZJ-7).

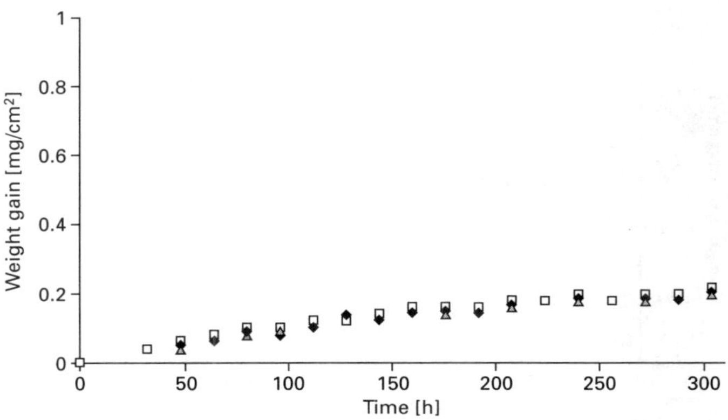

6.23 Net weight changes for steel AISI 441 during cyclic oxidation (8 h/4 h cycles) for 300 h at 800 °C in wet air (FZJ-8).

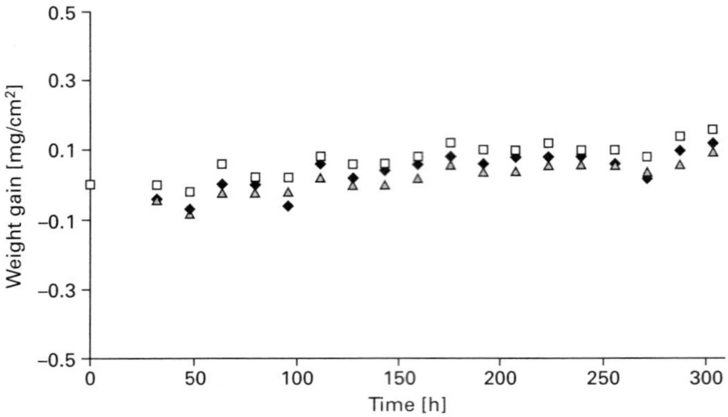

6.24 Gross weight changes for steel AISI 441 during cyclic oxidation (8h/4h cycles) for 300h at 800°C in wet air (FZJ-8).

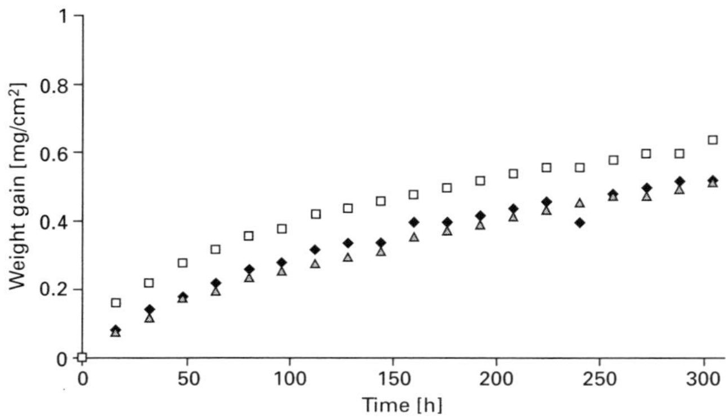

6.25 Net weight changes for steel AISI 441 during cyclic oxidation (8h/4h cycles) for 300h at 850°C in dry air (FZJ-9).

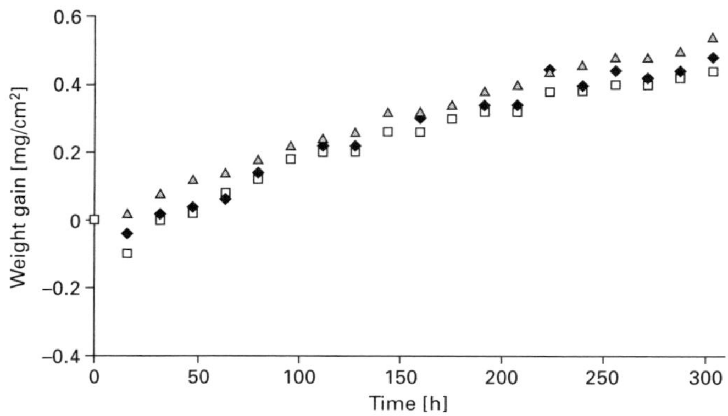

6.26 Gross weight changes for steel AISI 441 during cyclic oxidation (8h/4h cycles) for 300h at 850°C in dry air (FZJ-9).

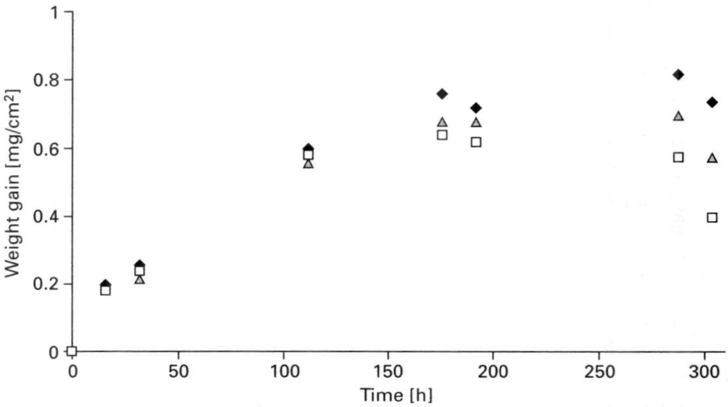

6.27 Net weight changes for steel AISI 441 during cyclic oxidation (8 h/2 h cycles) for 300 h at 900 °C in dry air (FZJ-10).

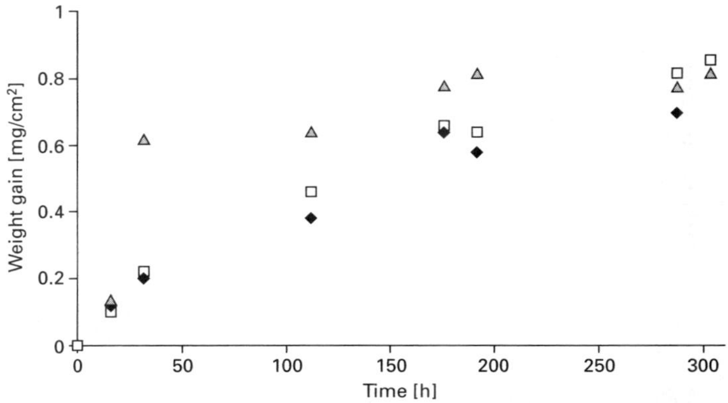

6.28 Gross weight changes for steel AISI 441 during cyclic oxidation (8 h/2 h cycles) for 300 h at 900 °C in dry air (FZJ-10).

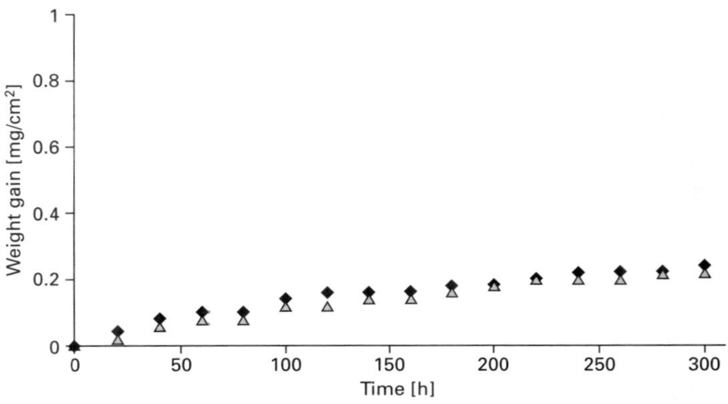

6.29 Net weight changes for steel AISI 441 during cyclic oxidation (20 h/4 h cycles) for 300 h at 800 °C in dry air (FZJ-11).

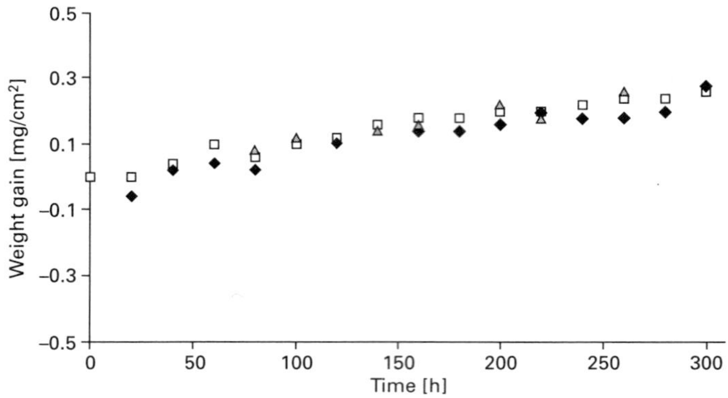

6.30 Gross weight changes for steel AISI 441 during cyclic oxidation (20 h/4 h cycles) for 300 h at 800 °C in dry air (FZJ-11).

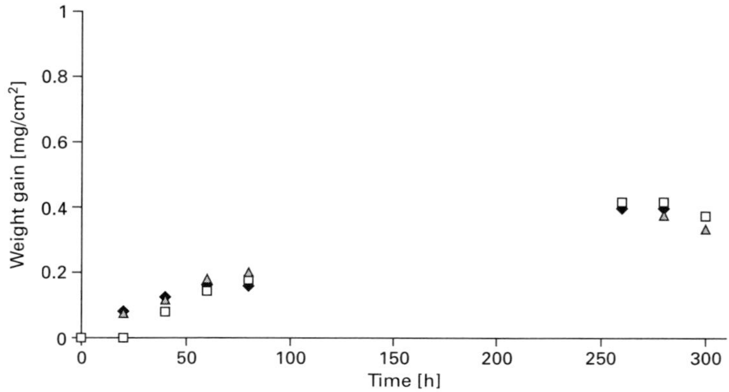

6.31 Net weight changes for steel AISI 441 during cyclic oxidation (20 h/2 h cycles) for 300 h at 850 °C in wet air (FZJ-12).

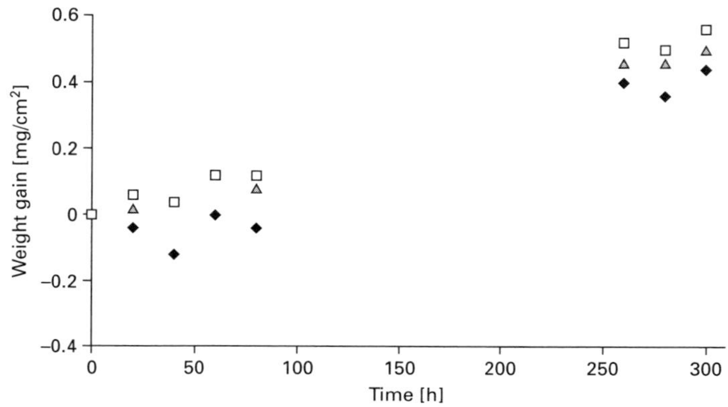

6.32 Gross weight changes for steel AISI 441 during cyclic oxidation (20 h/2 h cycles) for 300 h at 850 °C in wet air (FZJ-12).

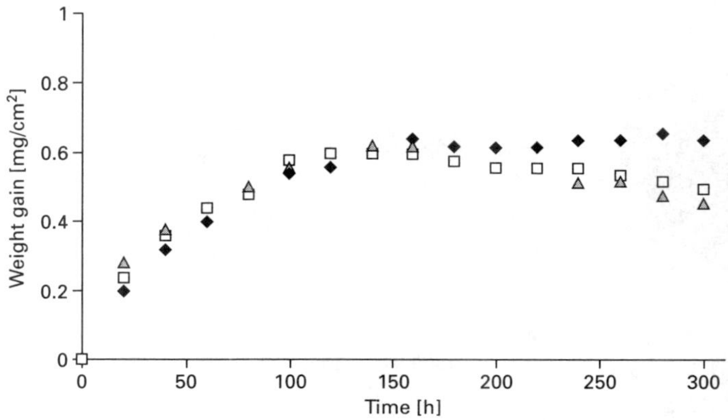

6.33 Net weight changes for steel AISI 441 during cyclic oxidation (20 h/4 h cycles) for 300 h at 900 °C in dry air (FZJ-13).

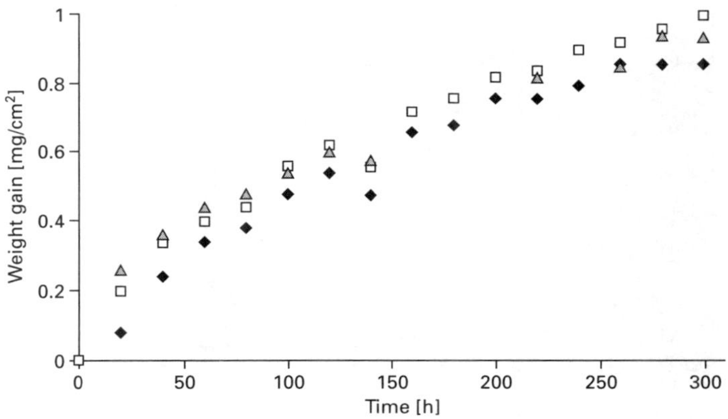

6.34 Gross weight changes for steel AISI 441 during cyclic oxidation (20 h/4 h cycles) for 300 h at 900 °C in dry air (FZJ-13).

behaviour was determined by near-parabolic oxide growth kinetics whereby only a minor difference between long and short dwell time testing was found. The onset of spallation occurred earlier for short dwell time testing than in the case of long dwell time testing. This effect was more pronounced at higher oxidation temperatures. The AISI 441 steel oxidised at temperatures in the range of 800–900 °C formed at its surface an oxide scale which consisted of an outer $(Mn, Cr)_3O_4$ layer, and an inner layer of Cr_2O_3 (Figs 6.35–6.44). The internal oxidation zone contained small precipitates of Al_2O_3 and TiO_2 (Figs 6.35 and 6.36). Scanning electron microscopy/energy dispersive X-ray (SEM/EDX) analysis of the specimen surfaces after oxidation showed the formation of 'nodules' containing TiO_2 (Figs 6.37 and 6.38). The number and average size of the 'nodules' increased with increasing oxidation temperature (Figs 6.37 and 6.38).

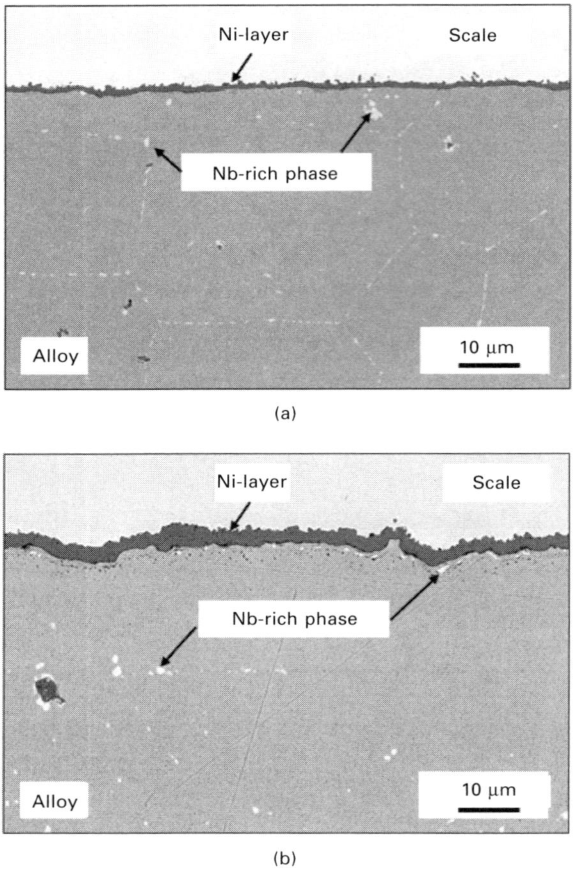

6.35 SEM/BSE image of microstructures of the oxide scales on steel AISI 441 after 300 h cyclic oxidation: (a) specimen oxidised at 800 °C, 8 h/4 h cycles in wet air; (b) specimen oxidised at 850 °C, 4 h/4 h cycles in dry air.

6.3.3 Oxidation of steel Alloy 800H

The net and gross mass change curves for Alloy 800H during exposure at 950, 1000 and 1050 °C are given in Figs 6.45–6.66. The early stages of exposure were determined by a near-parabolic time dependence of the scale growth rate. The time for the onset of spallation varied strongly as a function of temperature and upper dwell time and in general decreased by decreasing upper dwell time and increasing test temperatures. Shortening the upper dwell time and/or increasing the oxidation temperatures enhanced the extent of spallation.

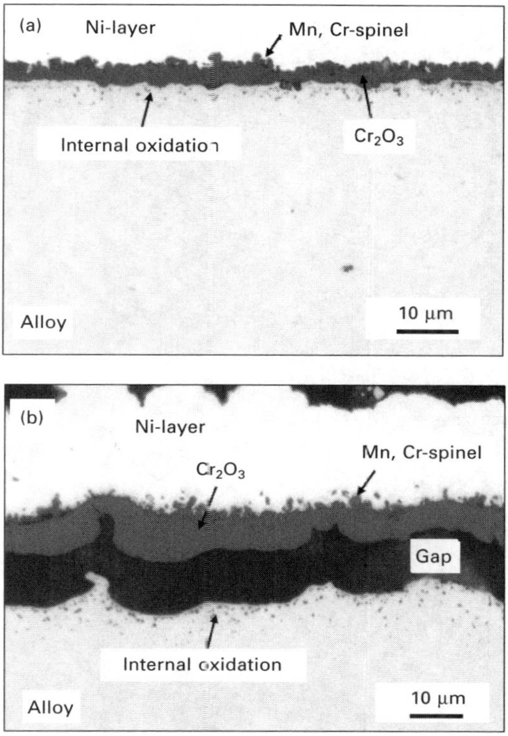

6.36 Microstructures of the oxide scales on steel AISI 441 after 300 h cyclic oxidation: (a) optical metallography image of specimen oxidised at 850 °C, 4 h/4 h cycles in dry air; (b) SEM/BSE image of the specimen oxidised at 900 °C, 4 h/4 h cycles in wet air.

Figures 6.67–6.70 show surface morphology and cross-sections of the oxide scales formed on Alloy 800H during cyclic oxidation at 950, 1000 and 1050 °C. During oxidation at 950 and 1000 °C the alloy formed at its surface an outer layer of $(Mn, Cr)_3O_4$ spinel and an inner layer of Cr_2O_3, as well as an internal oxidation zone (Figs 6.67–6.70). The substantial thermal expansion coefficient mismatch between metal matrix (austenite) and oxide leads, however, to spallation of the oxide scale. Increasing the oxidation temperature to 1050 °C resulted in enhanced spallation caused by significant changes of oxide scale morphology and phase composition (Figs 6.70–6.73). The oxide scale formed at 1050 °C was heterogeneous and consisted of Fe,Cr,Ni-spinel, NiO-type oxide and a chromia layer formed near the metal/oxide interface (Figs 6.70–6.73). Moreover, substantial porosity and void formation was observed in the internal oxidation zone.

6.37 SEM images of the oxide scale surfaces formed on steel AISI 441 during 300 h cyclic oxidation: (a) specimen oxidised at 850 °C, 4 h/4 h cycles in dry air; (b) specimen oxidised at 800 °C, 8 h/4 h in wet air.

6.3.4 Oxidation of alloy CM 247

The net and gross mass change curves for the Ni-based superalloy CM 247 are given in Figs 6.74–6.81. Contrary to the aforementioned materials, the varied parameters showed only limited (oxidation temperature) or virtually no (dwell times) influence on the oxidation/spallation kinetics of the alloy. The outer part of the oxide scales formed on the alloy surface spalled off during the first few cooling cycles. Increasing the oxidation temperature from 1100 to 1150 °C resulted in enhanced spallation during the first 24 h of oxidation. Figure 6.82 shows the cross-section of the oxide scale after cyclic oxidation for 300 h at 1100 °C. It consisted of an outer layer of $NiAl_2O_4$ and an inner layer of Al_2O_3 (Figs 6.82–6.84). In the internal oxidation zone precipitates of HfO_2 could be observed. The internal oxides were found at

6.38 SEM images of the oxide scale surfaces formed on steel AISI 441 during 300 h cyclic oxidation (4 h/4 h cycles) at 850 °C in dry air: (a) overview; (b) higher magnification of the outlined region in (a).

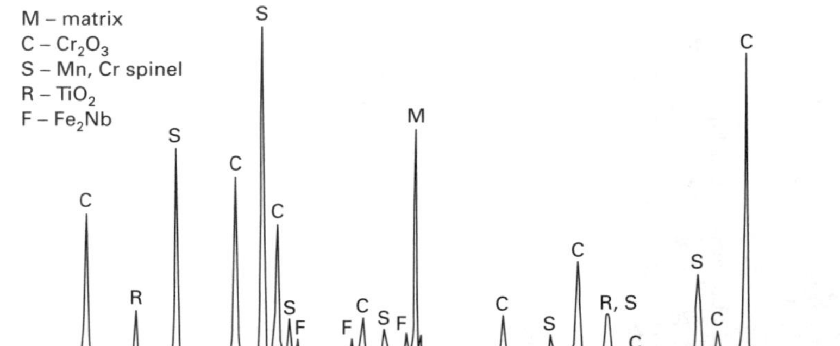

6.39 XRD pattern of oxide scale on steel AISI 441, after cyclic oxidation for 300 h (4 h/2 h cycles) at 800 °C in dry air.

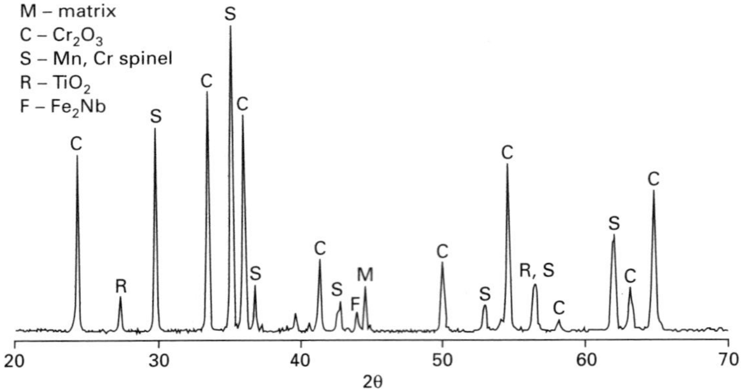

6.40 XRD pattern of oxide scale on steel AISI 441, after cyclic oxidation for 300h (8h/4h cycles) at 850°C in dry air.

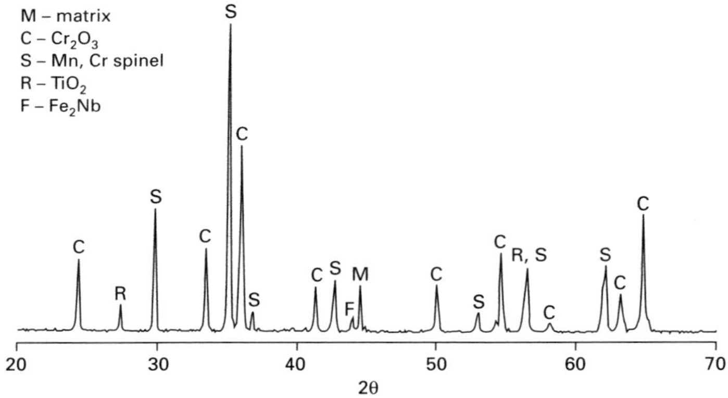

6.41 XRD pattern of oxide scale on steel AISI 441, after cyclic oxidation for 300h (4h/4h cycles) at 900°C in wet air.

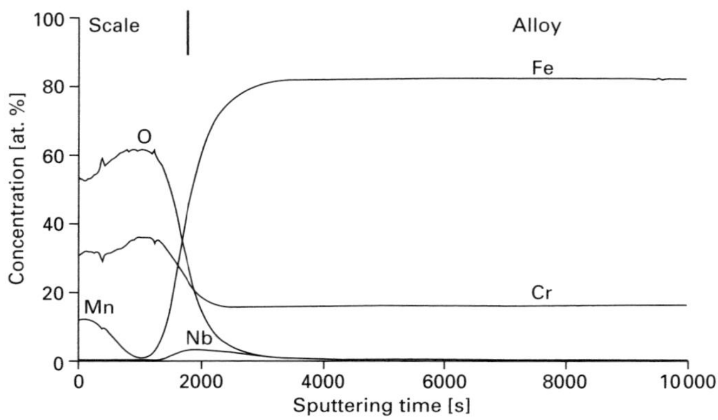

6.42 SNMS depth profile of steel AISI 441 after 300h cyclic oxidation (4h/2h cycles) at 800°C in dry air.

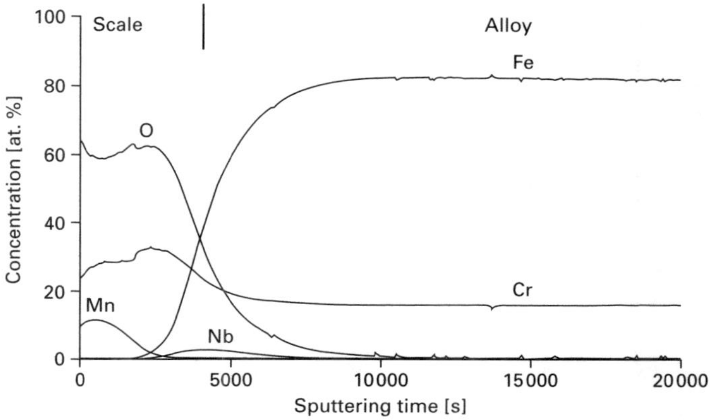

6.43 SNMS depth profile of steel AISI 441 after 300 h cyclic oxidation (4 h/4 h cycles) at 850 °C in dry air.

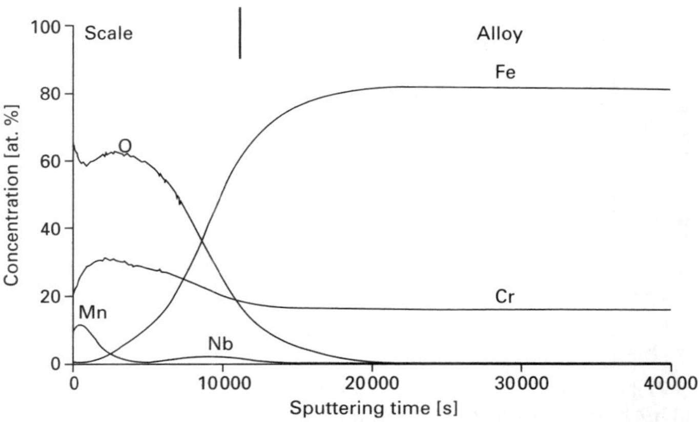

6.44 SNMS depth profile of steel AISI 441 after 300 h cyclic oxidation (4 h/4 h cycles) at 900 °C in wet air.

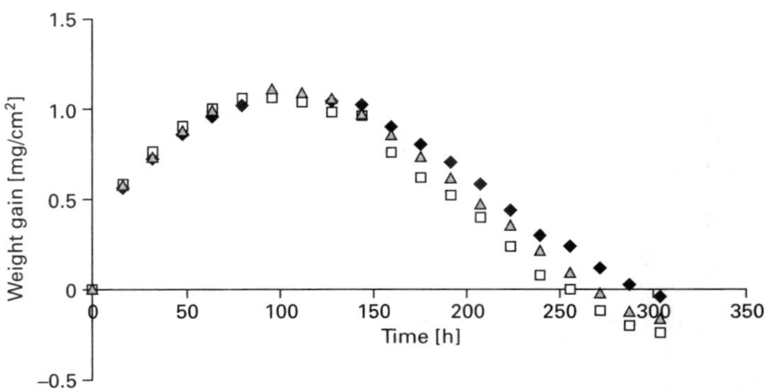

6.45 Net weight changes for steel Alloy 800H during cyclic oxidation (4 h/2 h cycles) for 300 h at 950 °C in dry air (FZJ-14).

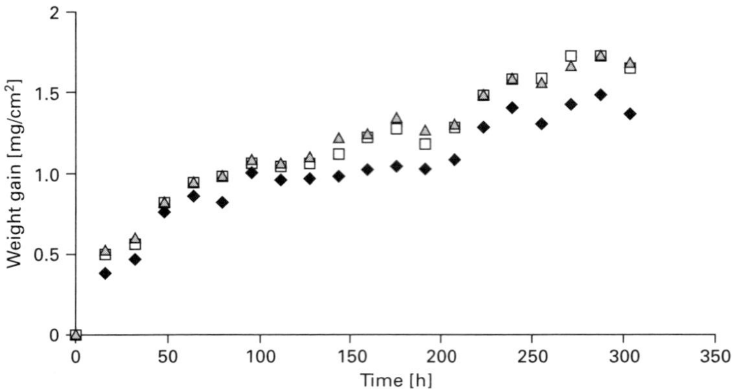

6.46 Gross weight changes for steel Alloy 800H during cyclic oxidation (4 h/2 h cycles) for 300 h at 950 °C in dry air (FZJ-14).

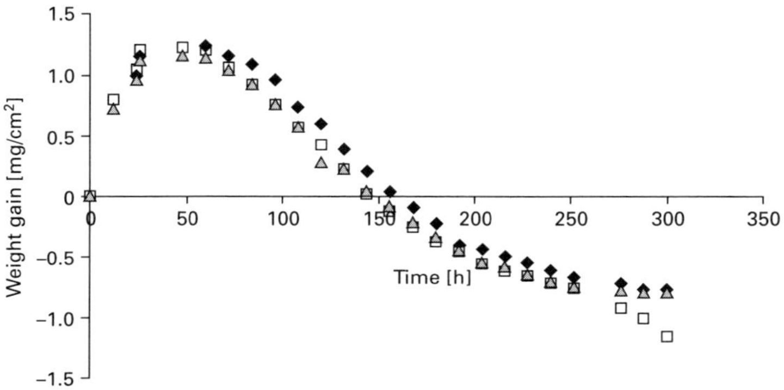

6.47 Net weight changes for steel Alloy 800H during cyclic oxidation (4 h/4 h cycles) for 300 h at 1000 °C in dry air (FZJ-15).

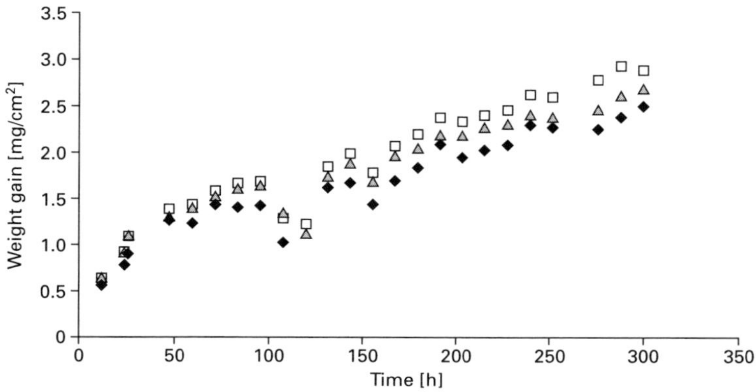

6.48 Gross weight changes for steel Alloy 800H during cyclic oxidation (4 h/4 h cycles) for 300 h at 1000 °C in dry air (FZJ-15).

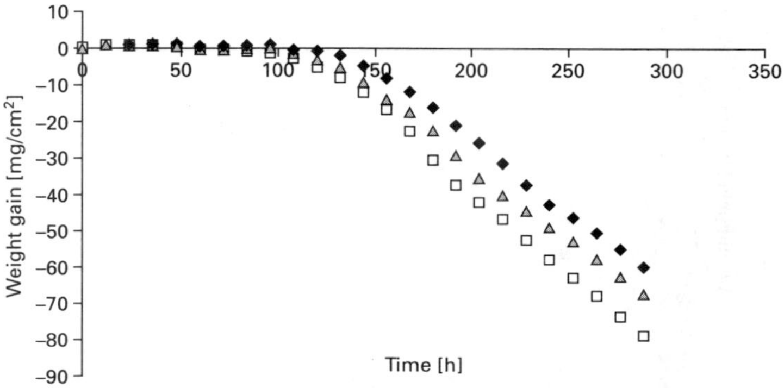

6.49 Net weight changes for steel Alloy 800H during cyclic oxidation (4 h/4 h cycles) for 300 h at 1050 °C in wet air (FZJ-16).

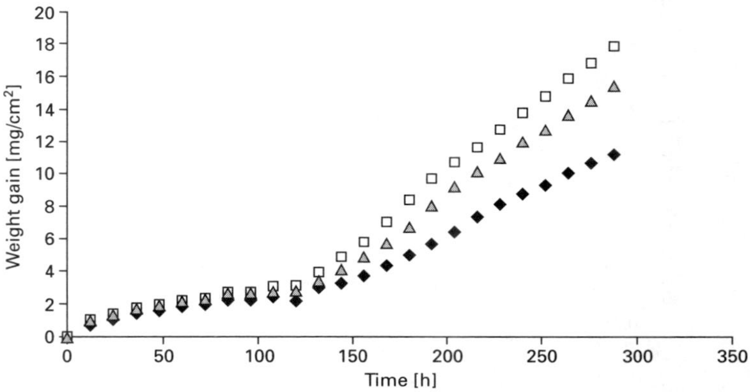

6.50 Gross weight changes for steel Alloy 800H during cyclic oxidation (4 h/4 h cycles) for 300 h at 1050 °C in wet air (FZJ-16).

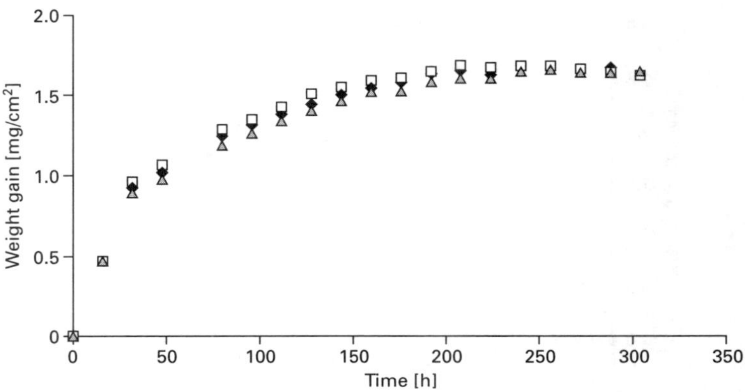

6.51 Net weight changes for steel Alloy 800H during cyclic oxidation (8 h/4 h cycles) for 300 h at 950 °C in wet air (FZJ-17).

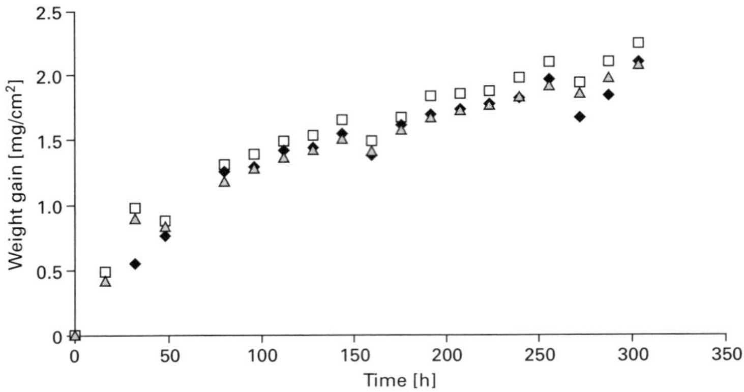

6.52 Gross weight changes for steel Alloy 800H during cyclic oxidation (8h/4h cycles) for 300h at 950°C in wet air (FZJ-17).

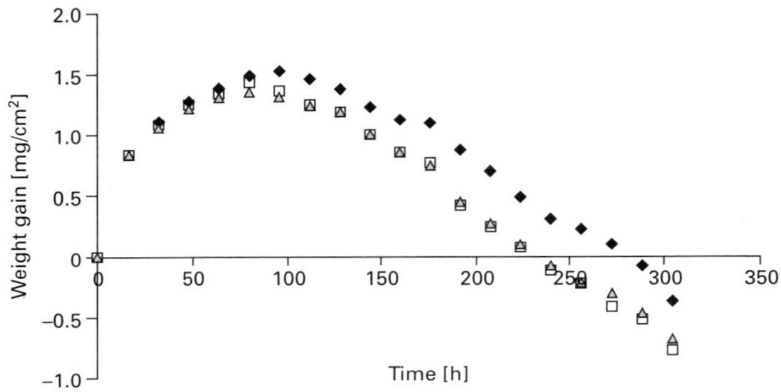

6.53 Net weight changes for steel Alloy 800H during cyclic oxidation (8h/4h cycles) for 300h at 1000°C in dry air (FZJ-18).

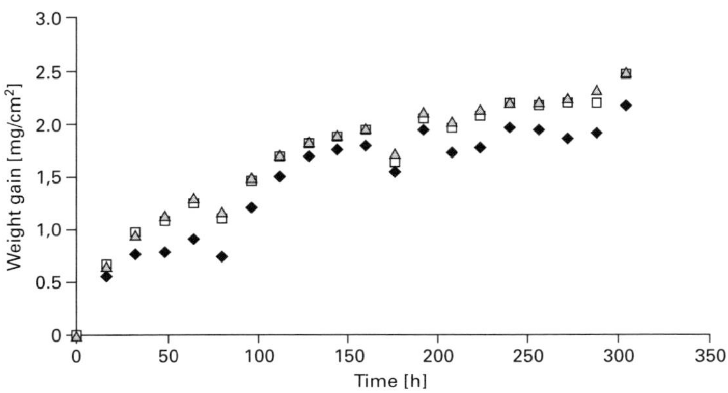

6.54 Gross weight changes for steel Alloy 800H during cyclic oxidation (8h/4h cycles) for 300h at 1000°C in dry air (FZJ-18).

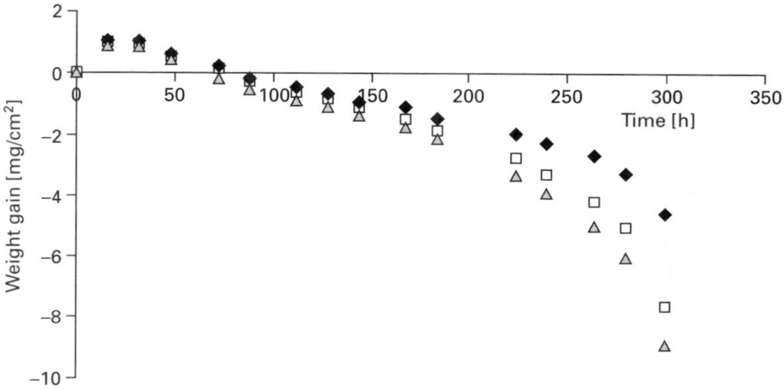

6.55 Net weight changes for steel Alloy 800H during cyclic oxidation (8 h/2 h cycles) for 300 h at 1050 °C in dry air (FZJ-19).

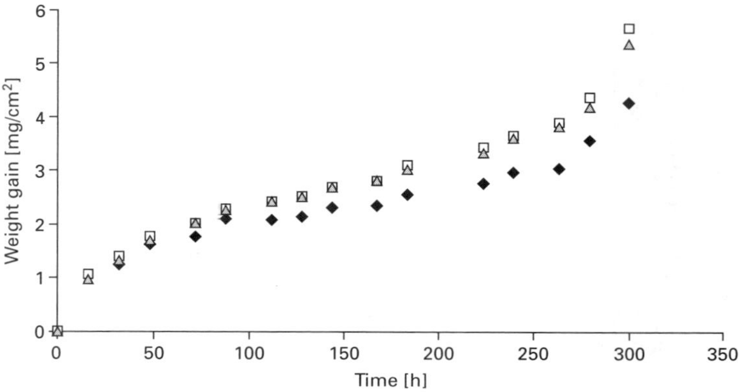

6.56 Gross weight changes for steel Alloy 800H during cyclic oxidation (8 h/2 h cycles) for 300 h at 1050 °C in dry air (FZJ-19).

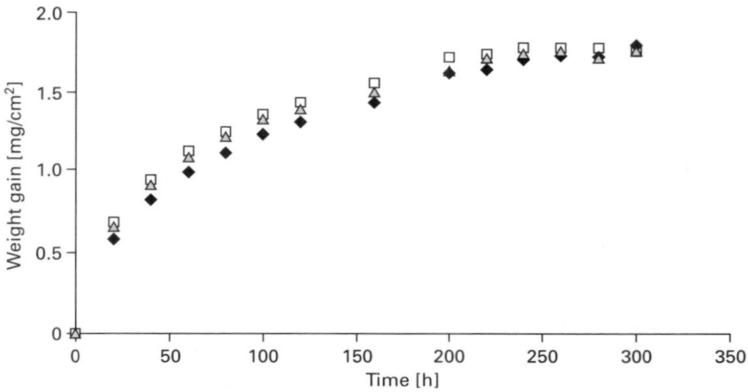

6.57 Net weight changes for steel Alloy 800H during cyclic oxidation (20 h/4 h cycles) for 300 h at 950 °C in dry air (FZJ-20).

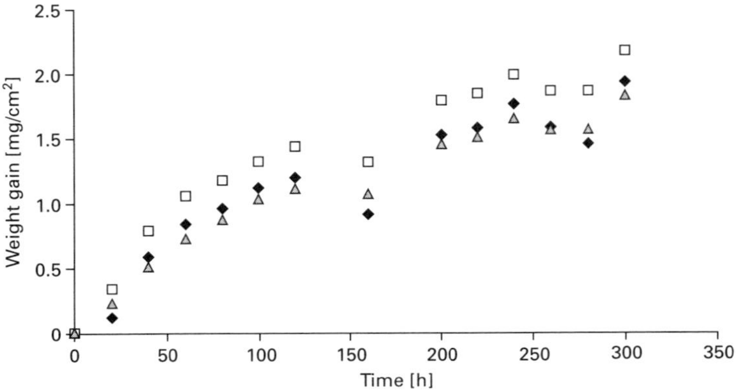

6.58 Gross weight changes for steel Alloy 800H during cyclic oxidation (20 h/4 h cycles) for 300 h at 950 °C in dry air (FZJ-20).

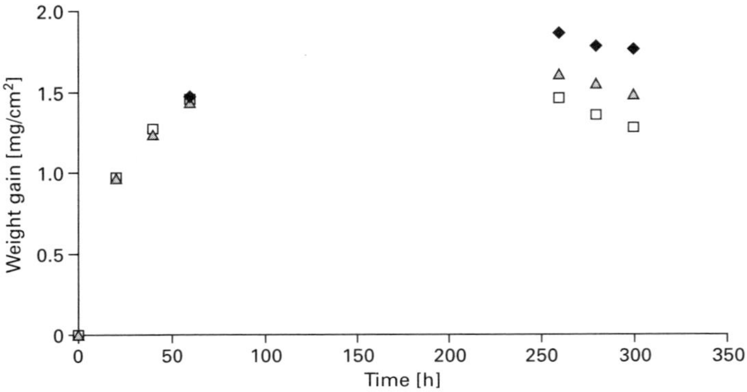

6.59 Net weight changes for steel Alloy 800H during cyclic oxidation (20 h/2 h cycles) for 300 h at 1000 °C in wet air (FZJ-21).

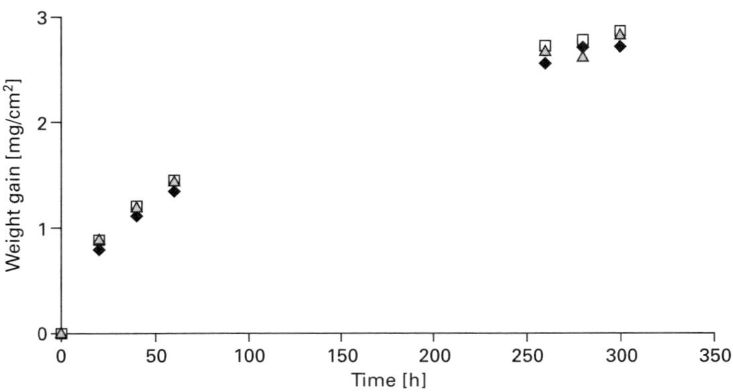

6.60 Gross weight changes for steel Alloy 800H during cyclic oxidation (20 h/2 h cycles) for 300 h at 1000 °C in wet air (FZJ-21).

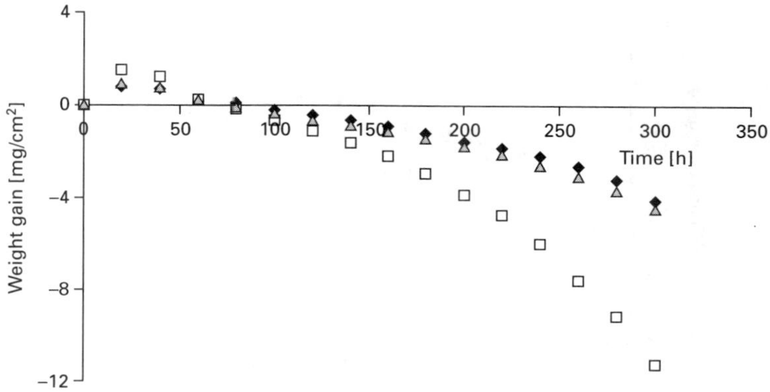

6.61 Net weight changes for steel Alloy 800H during cyclic oxidation (20 h/4 h cycles) for 300 h at 1050 °C in dry air (FZJ-22).

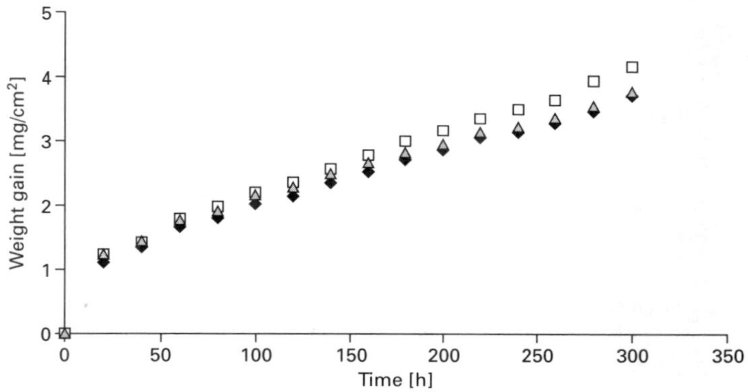

6.62 Gross weight changes for steel Alloy 800H during cyclic oxidation (20 h/4 h cycles) for 300 h at 1050 °C in dry air (FZJ-22).

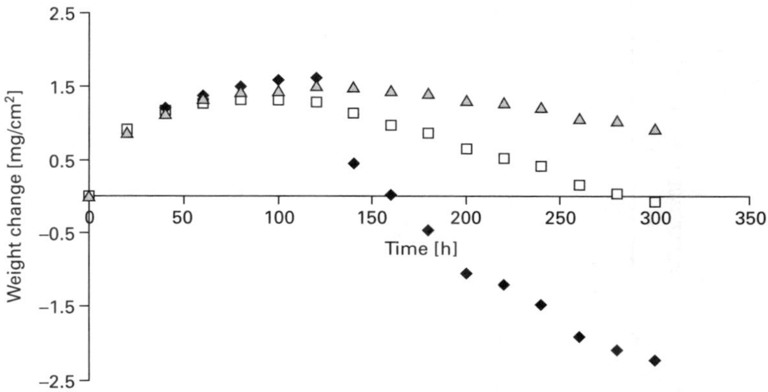

6.63 Net weight changes for steel Alloy 800H during cyclic oxidation (20 h/4 h cycles) for 300 h at 1000 °C in dry air (FZJ-22).

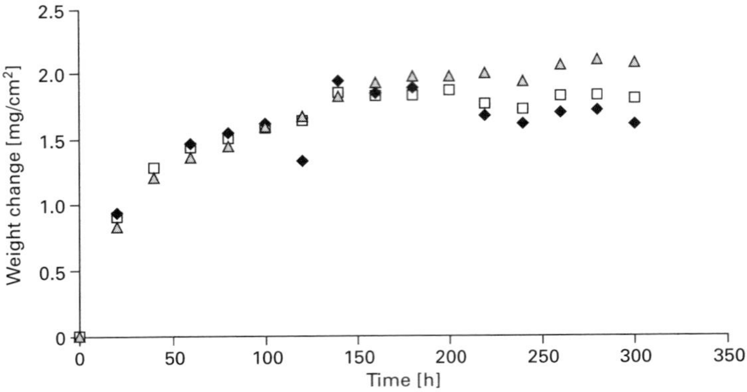

6.64 Gross weight changes for steel Alloy 800H during cyclic oxidation (20 h/4 h cycles) for 300 h at 1000 °C in dry air (FZJ-21).

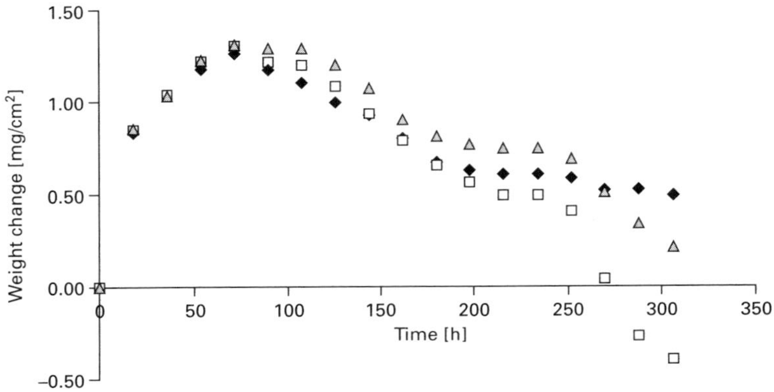

6.65 Net weight changes for steel Alloy 800H during cyclic oxidation (1 h/0.25 h cycles) for 300 h at 1000 °C in dry air (FZJ-33).

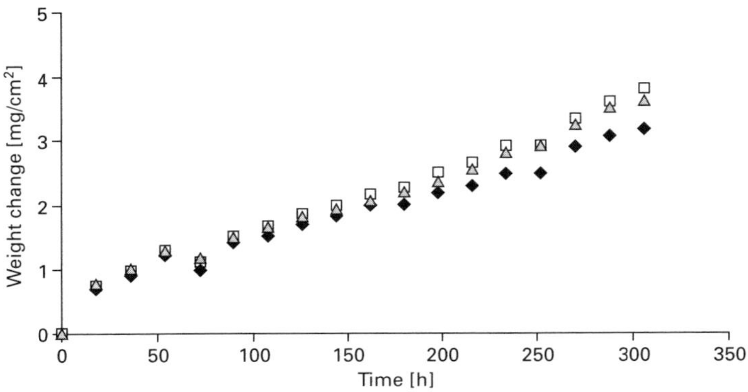

6.66 Gross weight changes for steel Alloy 800H during cyclic oxidation (1 h/0.25 h cycles) for 300 h at 1000 °C in dry air (FZJ-33).

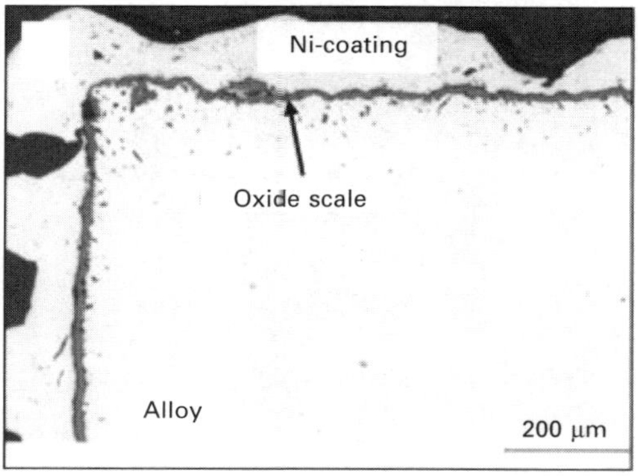

(a)

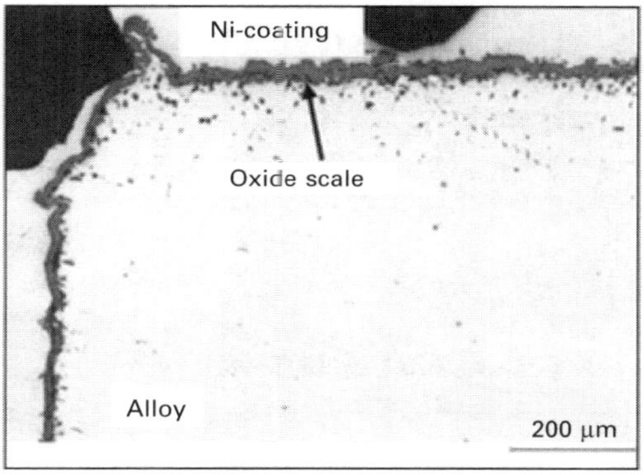

(b)

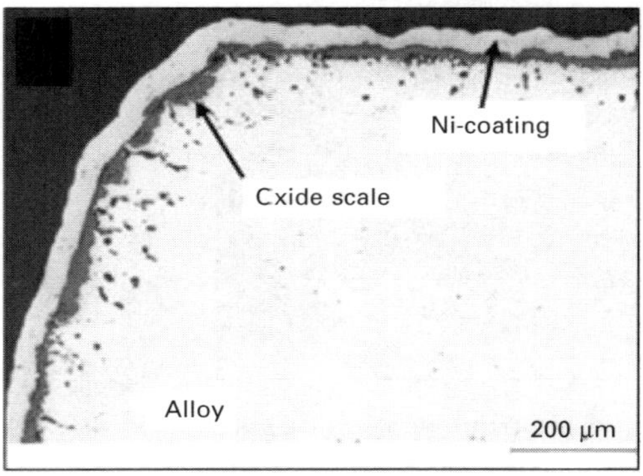

(c)

6.67 Optical metallography cross-sections of oxide scales formed on steel Alloy 800H during 300 h cyclic oxidation: (a) specimen oxidised at 950 °C, 8 h/4 h cycles in wet air; (b) specimen oxidised at 1000 °C, 4 h/4 h cycles in dry air; (c) specimen oxidised at 1050 °C, 8 h/2 h cycles in dry air.

6.68 SEM images of the oxide scale surfaces formed on steel Alloy 800H during 300 h cyclic oxidation: (a) specimen oxidised at 950 °C, 8 h/4 h cycles in wet air; (b) specimen oxidised at 1000 °C, 4 h/4 h cycles in dry air; (c) specimen oxidised at 1050 °C, 4 h/4 h cycles in wet air.

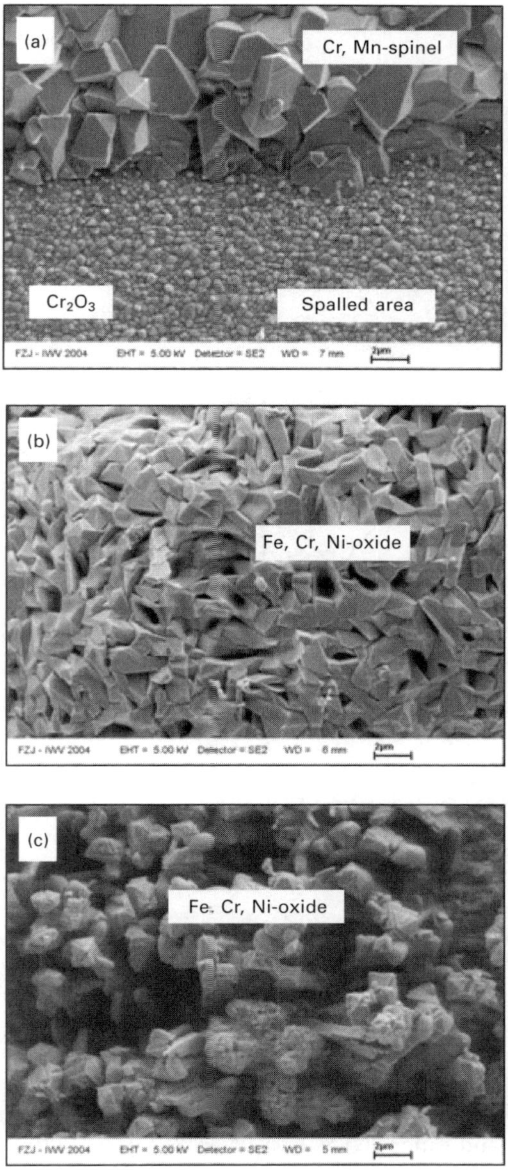

6.69 SEM images of the oxide scale surfaces formed on steel Alloy 800H during 300 h cyclic oxidation: (a) specimen oxidised at 950 °C, 8 h/4 h cycles in wet air; (b) specimen oxidised at 1000 °C, 4 h/4 h cycles in dry air; (c) specimen oxidised at 1050 °C, 4 h/4 h cycles in wet air.

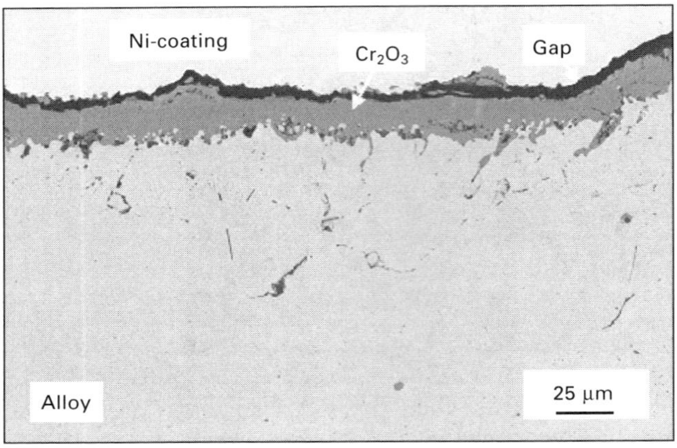

(a)

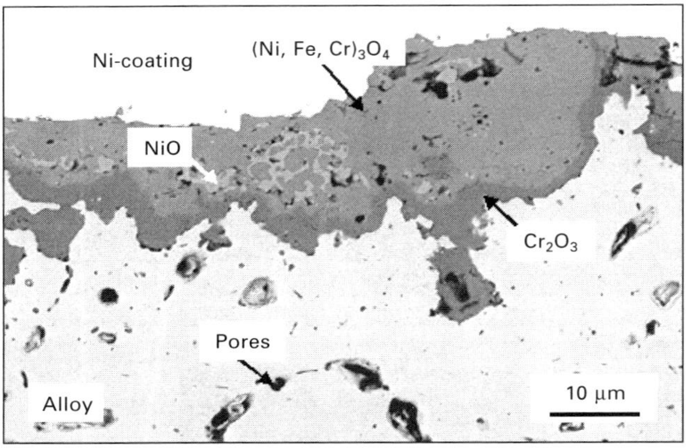

(b)

6.70 SEM/BSE image of microstructures of the oxide scales on steel Alloy 800H after 300 h cyclic oxidation: (a) specimen oxidised at 1000 °C, 20 h/4 h cycles in wet air; (b) specimen oxidised at 1050 °C, 4 h/4 h cycles in wet air.

the alloy grain boundaries, i.e. they were mainly formed out of the initially prevailing carbides. The oxide spallation tended to occur at the interface between inner (Al_2O_3) and outer ($NiAl_2O_4$) layer. After spallation of large parts of the outer scale, scale growth is determined by transport processes in the inner, dense alumina scale and thus possesses a substantially lower growth rate than that during or just before the occurrence of spalling.

6.3.5 Oxidation of alloy Kanthal A1

Figure 6.85 shows the design of the test facility specially set up for the

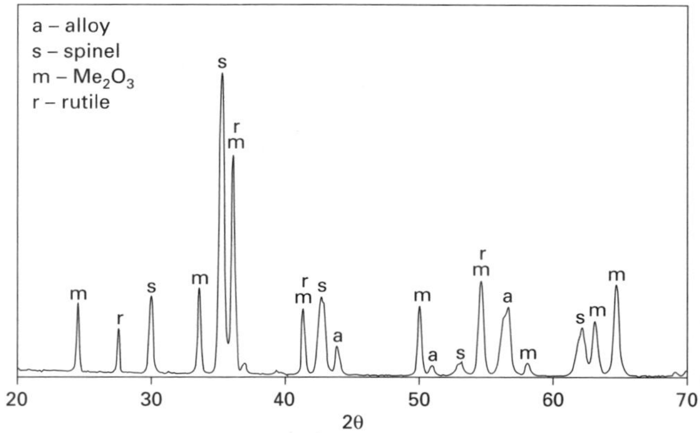

6.71 XRD pattern of oxide scale on steel Alloy 800H, after cyclic oxidation for 300 h (8 h/4 h cycles) at 950 °C in wet air.

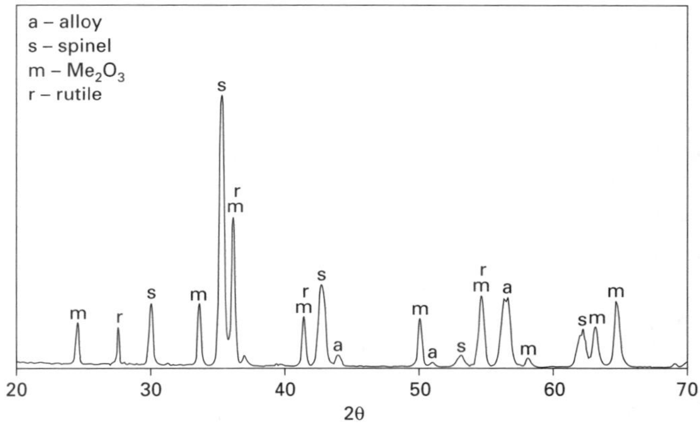

6.72 XRD pattern of oxide scale on steel Alloy 800H, after cyclic oxidation for 300 h (20 h/2 h cycles) at 1000 °C in wet air.

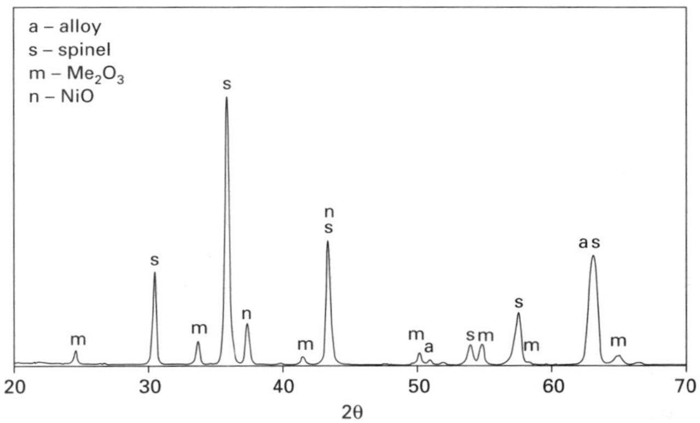

6.73 XRD pattern of oxide scale on steel Alloy 800H, after cyclic oxidation for 300 h (4 h/4 h cycles) at 1050 °C in wet air.

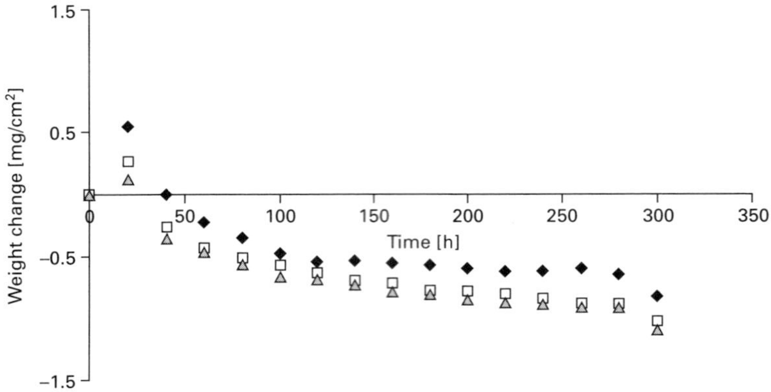

6.74 Net weight changes for alloy CM 247 during cyclic oxidation (20 h/4 h cycles) for 300 h at 1100 °C in dry air (FZJ-23).

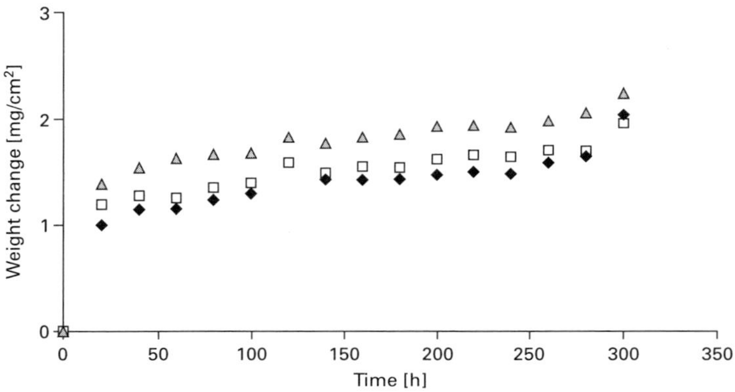

6.75 Gross weight changes for alloy CM 247 during cyclic oxidation (20 h/4 h cycles) for 300 h at 1100 °C in dry air (FZJ-23).

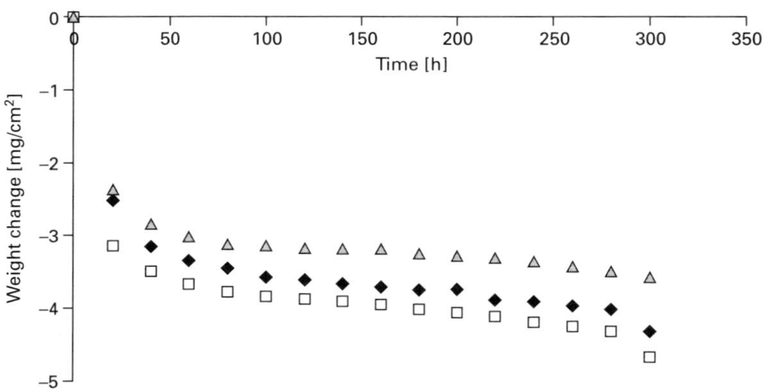

6.76 Net weight changes for alloy CM 247 during cyclic oxidation (20 h/4 h cycles) for 300 h at 1150 °C in dry air (FZJ-24).

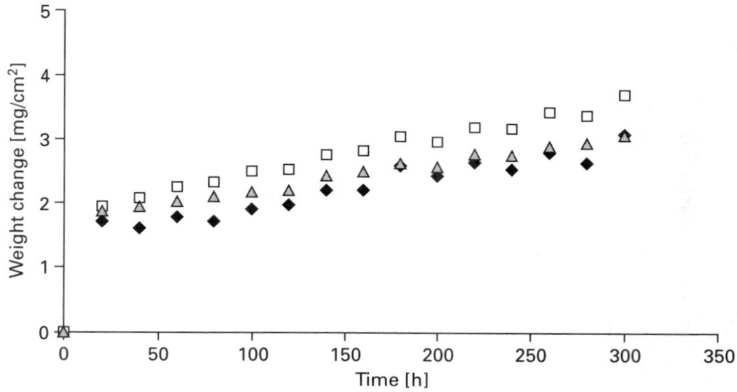

6.77 Gross weight changes for alloy CM 247 during cyclic oxidation (20h/4h cycles) for 300h at 1150°C in dry air (FZJ-24).

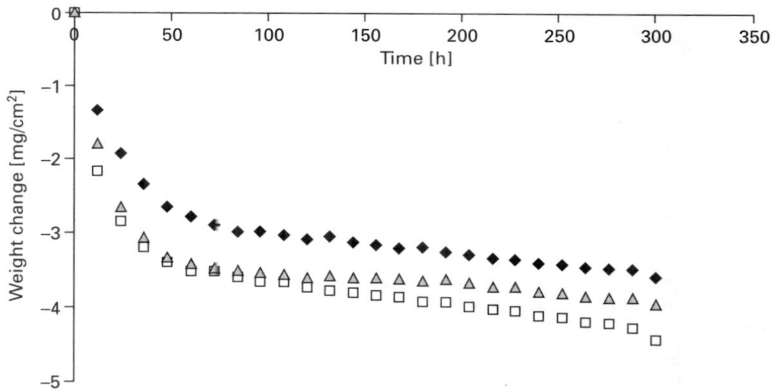

6.78 Net weight changes for alloy CM 247 during cyclic oxidation (4h/4h cycles) for 300h at 1150°C in dry air (FZJ-25).

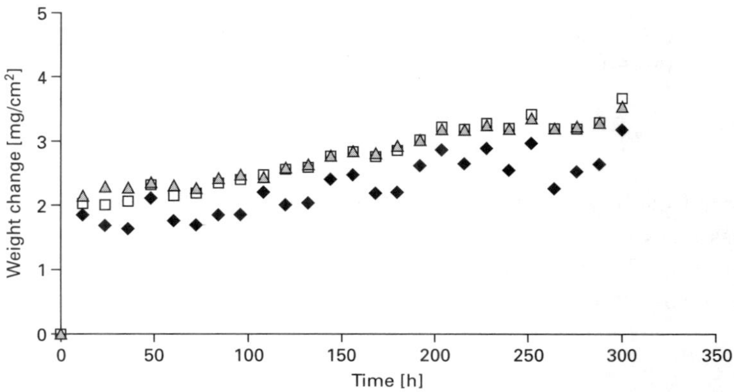

6.79 Gross weight changes for alloy CM 247 during cyclic oxidation (4h/4h cycles) for 300h at 1150°C in dry air (FZJ-25).

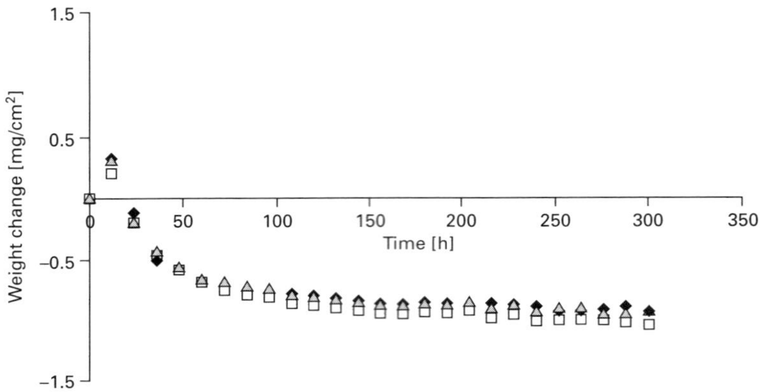

6.80 Net weight changes for alloy CM 247 during cyclic oxidation (4h/4h cycles) for 300h at 1100 °C in dry air (FZJ-26).

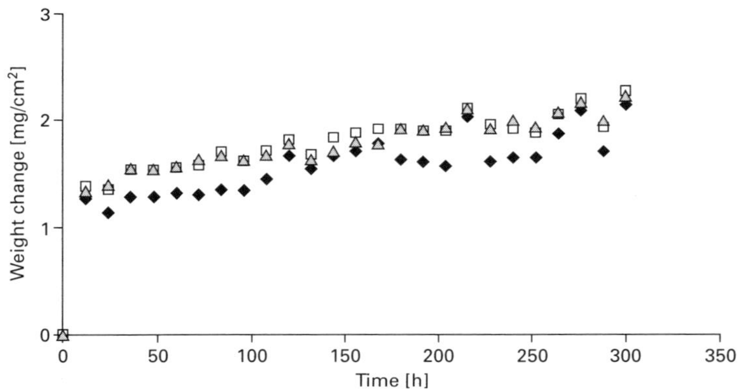

6.81 Gross weight changes for alloy CM 247 during cyclic oxidation (4h/4h cycles) for 300h at 1100 °C in dry air (FZJ-26).

thermal cycling oxidation testing of Kanthal A1 at 1250 °C. The net and gross mass change curves for the Kanthal A1 are given in Figs 6.86–6.89. Figures 6.90 and 6.91 show the oxide scales formed during cyclic oxidation in a macroscopic and two cross-sectional views.

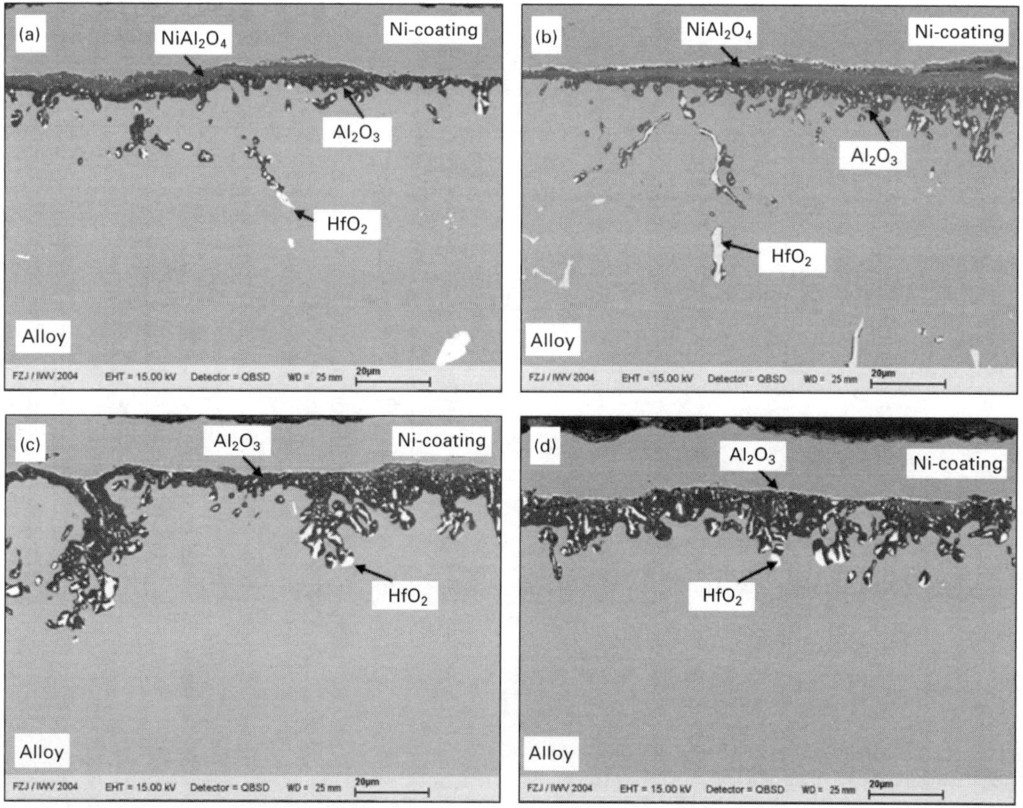

6.82 SEM/BSE images of microstructures of the oxide scales formed on alloy CM 247 during 300 h cyclic oxidation in dry air: (a) specimen oxidised at 1000 °C, 4 h/4 h cycles; (b) specimen oxidised at 1100 °C, 20 h/4 h cycles; (c) specimen oxidised at 1150 °C, 4 h/4 h cycles; (d) specimen oxidised at 1150 °C, 20 h/4 h cycles.

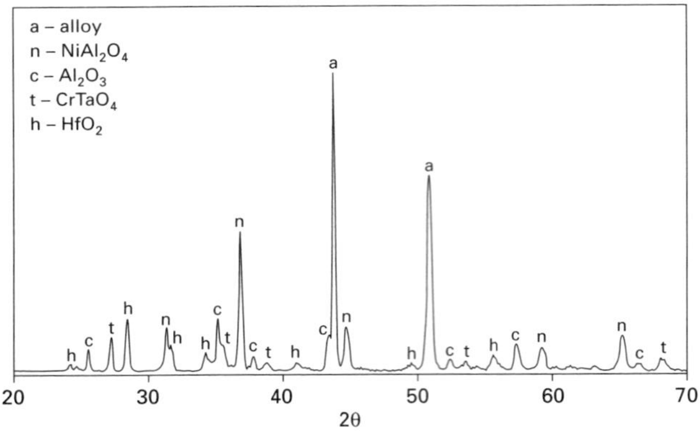

6.83 XRD pattern of oxide scale on alloy CM 247, after cyclic oxidation for 300 h (4 h/4 h cycles) at 1100 °C in dry air.

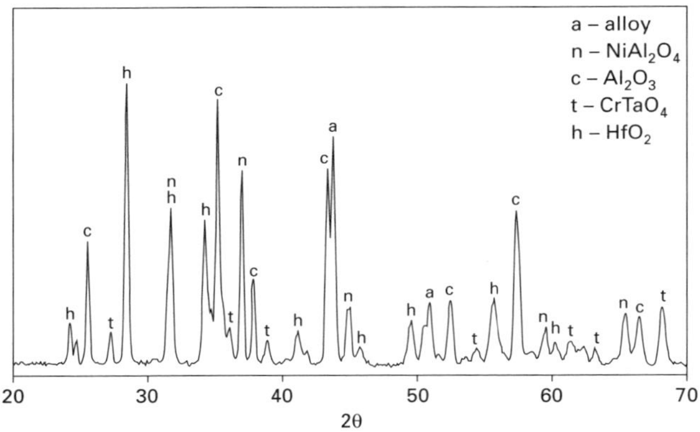

6.84 XRD pattern of oxide scale on alloy CM 247, after cyclic oxidation for 300 h (20 h/4 h cycles) at 1150 °C in dry air.

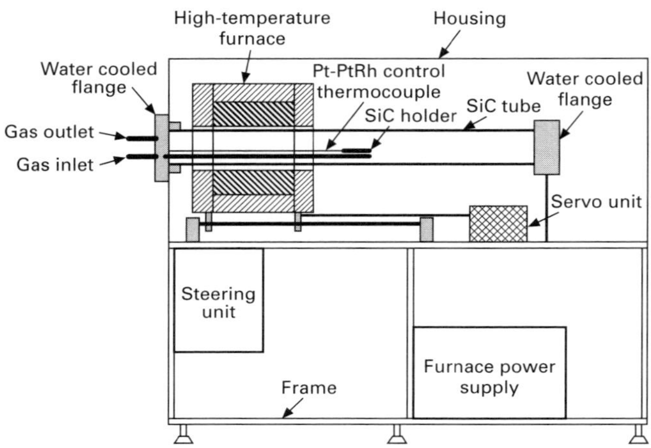

6.85 Facility for thermal cyclic oxidation of Kanthal A1 at 1250 °C.

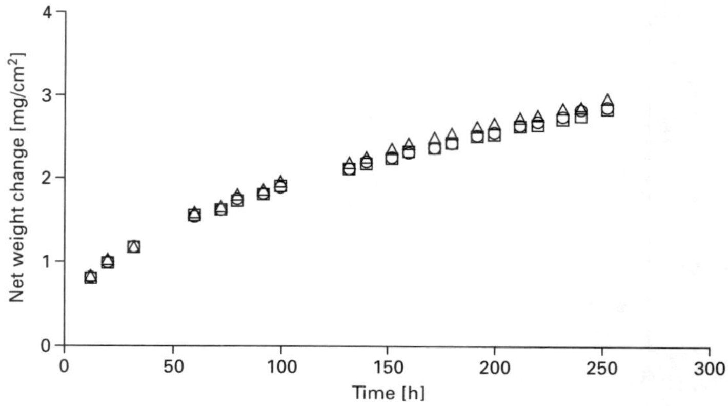

6.86 Net mass change of Kanthal A1 specimens during cyclic oxidation for 300 h, at 1250 °C in dry air (4 h/4 h cycles).

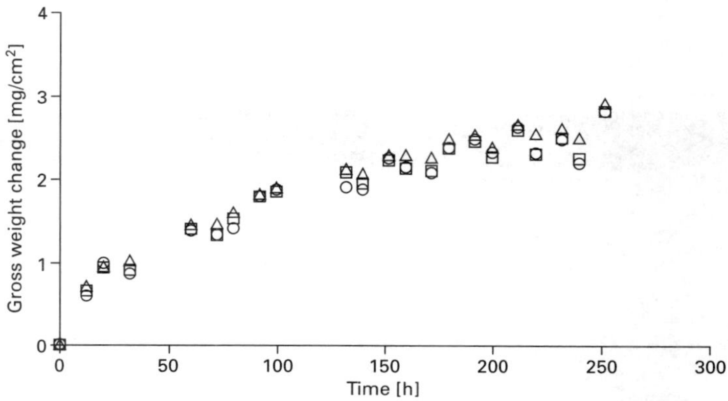

6.87 Gross mass change of Kanthal A1 specimens during cyclic oxidation for 300 h, at 1250 °C in dry air (4 h/4 h cycles).

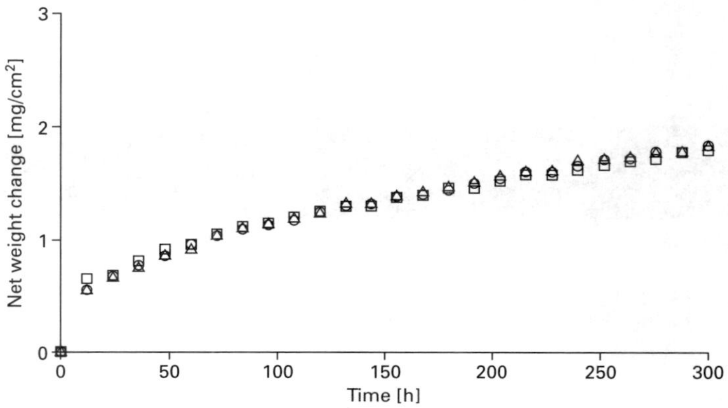

6.88 Net mass change of Kanthal A1 specimens during cyclic oxidation for 300 h, at 1200 °C in dry air (4 h/4 h cycles).

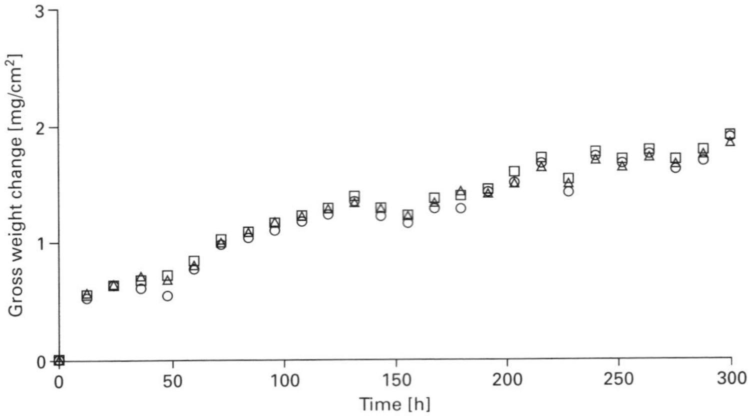

6.89 Gross mass change of Kanthal A1 specimens during cyclic oxidation for 300 h, at 1200 °C in dry air (4 h/4 h cycles).

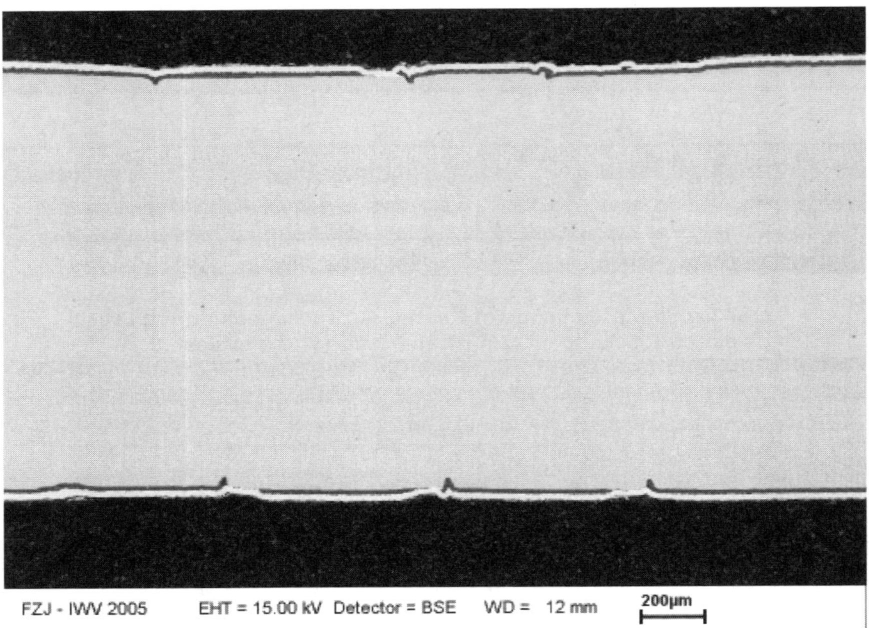

6.90 Cross-section of the oxide scale formed on alloy Kanthal A1 during 300 h cyclic oxidation at 1250 °C in dry air (4 h/4 h cycles).

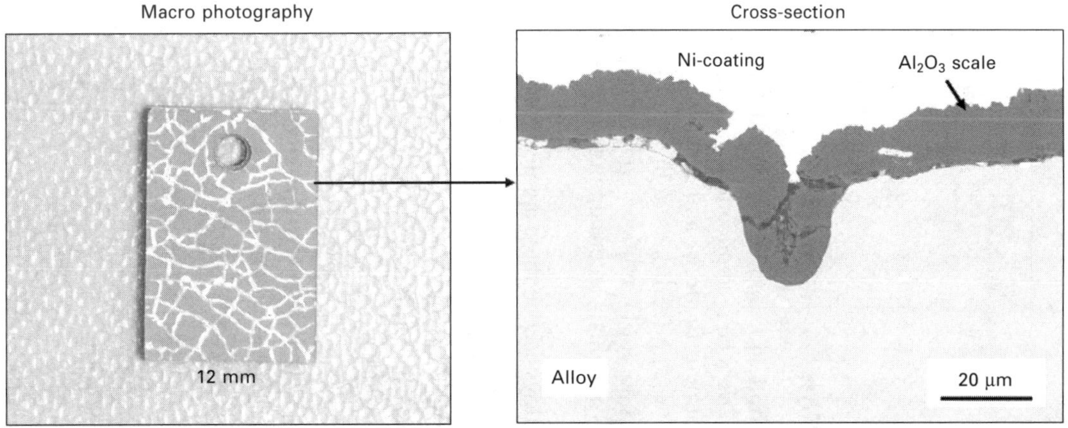

6.91 Oxide scale formed on Kanthal A1 during 300 h cyclic oxidation at 1250 °C in dry air (4 h/4 h cycles).

7

Investigation of the influence of parameter variation in short dwell thermal cycling oxidation

M. SCHÜTZE and M. MALESSA,
DECHEMA e.V., Germany

7.1 Introduction

Short dwell time tests find wide application in industry. This type of thermal cycling is typically experienced in applications such as industrial gas turbines, aero engines, heat treatment facilities and furnaces. The intervals between start-up and shut-down of the facilities are in the order of a few hours. The design life and/or the time until complete overhaul/repair is typically in the range of 10000–30000h with an accordant high number of several thousand temperature cycles.

The lifetime of the alloys used in these applications is decisively determined by their behaviour under cyclic oxidation conditions which is why particularly manufacturers and users of these alloys have great interest in standardising the cyclic oxidation test. Thus, the major objective of the 'short dwell time test standardisation' was to define a test method with wide applicability to the relevant industrial sectors.

7.2 Experimental investigation of reference materials under internally standardised thermal cycling oxidation conditions

7.2.1 Short dwell thermal cycling exposure test matrix

The test duration for each trial had to be extended to an accumulated length of the hot dwell time of 300h, as the originally proposed duration of 100h turned out to be too short to allow reliable correlations between the cycling testing parameters and cyclic oxidation behaviour. Therefore the matrix initially planned had to be adapted to fit into the time frame of the project and to maintain maximum output from the experiments as written in the technical report of the second period. The concept of a statistical balanced design was followed for every submatrix of different materials. Each test used three replicates. The matrix is given in Table 7.1.

Table 7.1 Test matrix for short dwell cyclic testing

Material	Test	Test conditions
AISI 441	F409_11x	5A2.1: 0.5 h hot, 1 h cold, 800 °C, dry
AISI 441	F409_12x	5A2.2: 0.5 h hot, 0.25 h cold, 850 °C, dry
AISI 441	F409_13x	5A2.3: 0.5 h hot, 0.25 h cold, 900 °C, wet
AISI 441	F409_14x	5A2.4: 1 h hot, 0.25 h cold, 800 °C, wet
AISI 441	F409_15x	5A2.5: 1 h hot, 1 h cold, 850 °C, dry
AISI 441	F409_16x	5A2.6: 1 h hot, 0.25 h cold, 900 °C, dry
AISI 441	F409_17x	5A2.7: 2 h hot, 0.25 h cold, 800 °C, dry
AISI 441	F409_18x	5A2.8: 2 h hot, 1 h cold, 850 °C, wet
AISI 441	F409_19x	5A2.9: 2 h hot, 0.25 h cold, 900 °C, dry
Alloy 800H	F409_21x	5A2.1: 0.5 h hot, 1 h cold, 950 °C, dry
Alloy 800H	F409_22x	5A2.2: 0.5 h hot, 0.25 h cold, 1000 °C, dry
Alloy 800H	F409_23x	5A2.3: 0.5 h hot, 0.25 h cold, 1050 °C, wet
Alloy 800H	F409_24x	5A2.4: 1 h hot, 0.25 h cold, 950 °C, wet
Alloy 800H	F409_25x	5A2.5: 1 h hot, 0.25 h cold, 1000 °C, dry
Alloy 800H	F409_26x	5A2.6: 1 h hot, 1 h cold, 1050 °C, dry
Alloy 800H	F409_27x	5A2.7: 2 h hot, 0.25 h cold, 950 °C, dry
Alloy 800H	F409_28x	5A2.8: 2 h hot, 1 h cold, 1000 °C, wet
Alloy 800H	F409_29x	5A2.9: 2 h hot, 0.25 h cold, 1050 °C, dry
CM 247	F409_31x	5A2.1: 1 h hot, 0.25 h cold, 1100 °C, dry
CM 247	F409_32x	5A2.2: 2 h hot, 0.25 h cold, 1150 °C, dry
CM 247	F409_33x	5A2.3: 1 h hot, 0.25 h cold, 1150 °C, dry
CM 247	F409_34x	5A2.4: 2 h hot, 0.25 h cold, 1100 °C, dry
Kanthal A1	F409_41x	5A2.1: 1 h hot, 0.25 h cold, 1200 °C, lab air
Kanthal A1	F409_42x	5A2.2: 2 h hot, 0.25 h cold, 1250 °C, lab air
Kanthal A1	F409_43x	5A2.3: 1 h hot, 0.25 h cold, 1250 °C, lab air
Kanthal A1	F409_44x	5A2.4: 2 h hot, 0.25 h cold, 1200 °C, lab air
P91	F409_51x	5A2.1: 1 h hot, 0.25 h cold, 650 °C, dry
P91	F409_52x	5A2.2: 2 h hot, 0.25 h cold, 650 °C, wet
P91	F409_53x	5A2.3: 1 h hot, 0.25 h cold, 650 °C, wet
P91	F409_54x	5A2.4: 2 h hot, 0.25 h cold, 650 °C, dry

7.2.2 Results of short dwell thermal cycling oxidation experiments

The net mass change of the test pieces as a function of accumulated exposure time is the standard way of representation of experimental isothermal and cyclic oxidation data. In Figs 7.1 to 7.9 the net mass change for the various materials is shown. Each figure represents one submatrix with parameter variation at a single temperature.

It is obvious that the choice of parameters influences the oxidation behaviour of the samples in a more or less pronounced way. While initial oxidation behaviour (growth of the oxide scales) seems to be independent of the selected parameters as revealed by statistical evaluation, the onset and rate of spallation depends significantly on the choice of the cycling parameters. Statistical

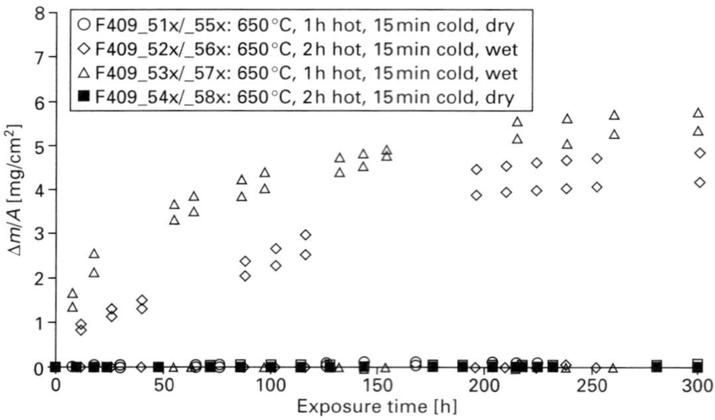

7.1 Net mass change of P91 at 650 °C (open symbols represent heat-treated samples).

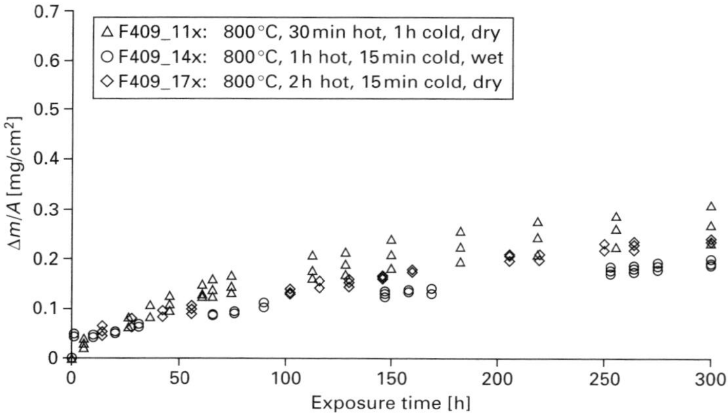

7.2 Net mass change of AISI 441 at 800 °C.

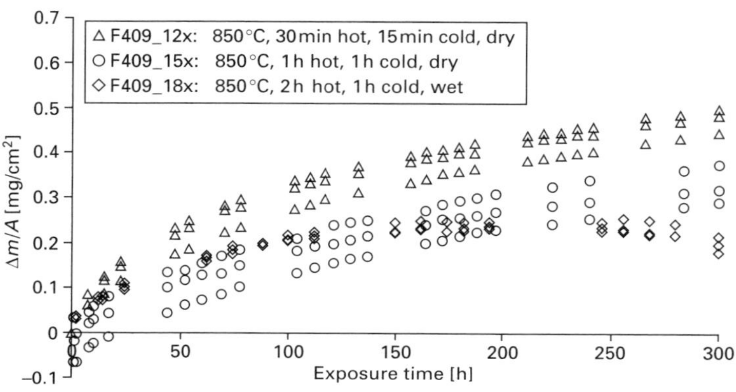

7.3 Net mass change of AISI 441 at 850 °C.

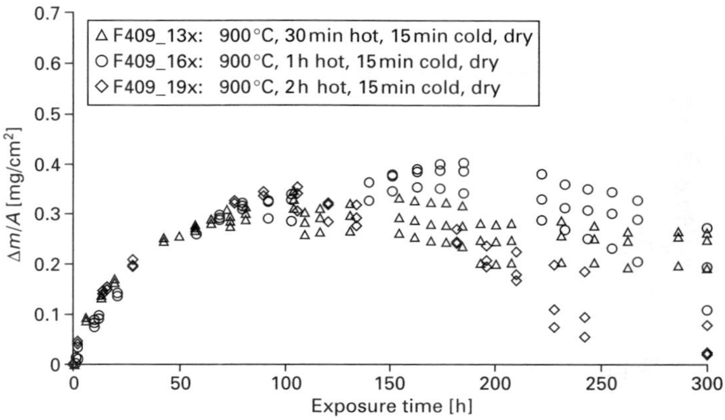

7.4 Net mass change of AISI 441 at 900°C.

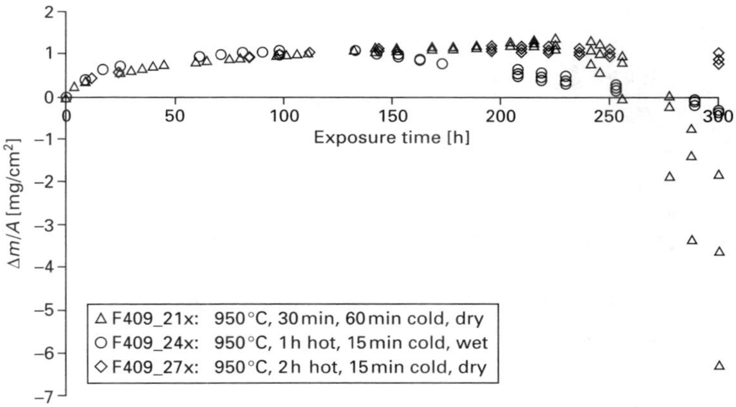

7.5 Net mass change of Alloy 800 H at 950°C.

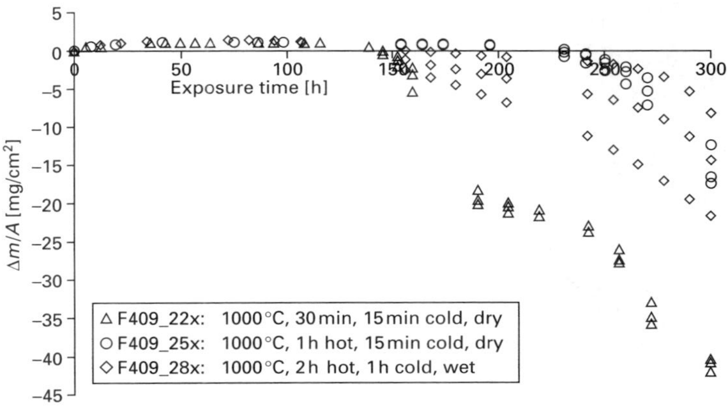

7.6 Net mass change of Alloy 800 H at 1000°C.

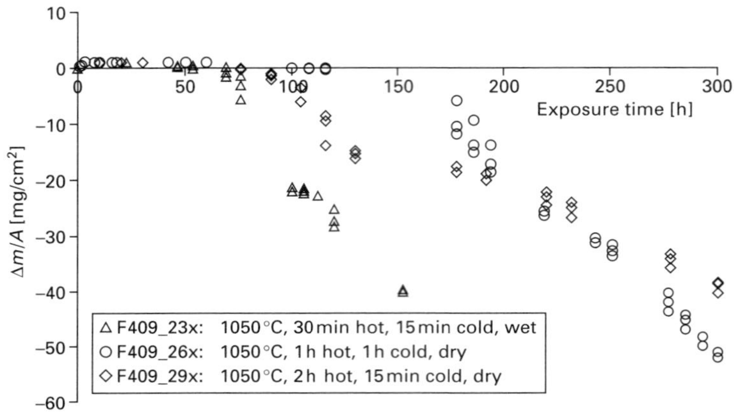

7.7 Net mass change of Alloy 800 H at 1050 °C.

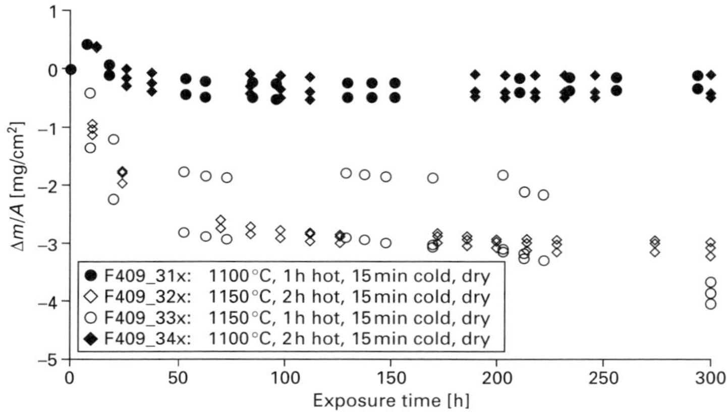

7.8 Net mass change of CM 247 at 1100 and 1150 °C.

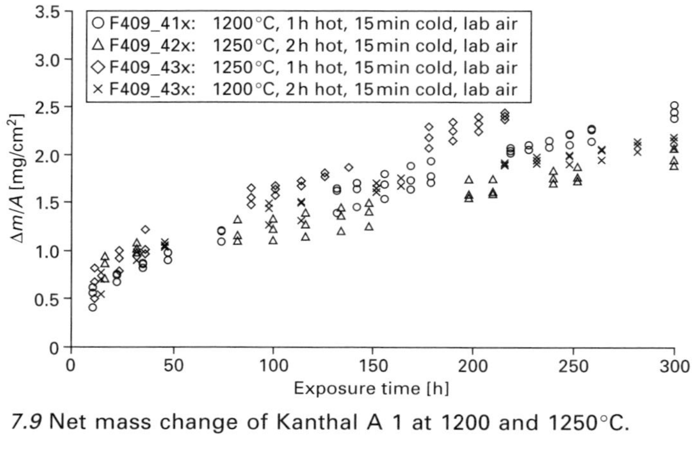

7.9 Net mass change of Kanthal A 1 at 1200 and 1250 °C.

evaluation on completed submatrices has been performed by University of Newcastle upon Tyne and revealed the most significant parameters.

For the ferritic steel P91 significant net mass changes were observed solely for the heat-treated specimens in humidified air. The non-heat-treated P91 specimens oxidised in dry and wet air as well as the heat-treated specimens oxidised in dry air showed only very limited mass change (Fig. 7.1). These negligible mass changes result from the formation of an external thin, protective oxide scale during the cyclic testing. In contrast, an enhanced oxidation as the result of a breakaway process was observed for the heat-treated material in wet air. The non-protective scale consisted of an outer layer of haematite, inner layers of Fe_3O_4 (magnetite) and $(Fe, Cr)_3O_4$ spinel and a narrow internal oxidation zone with chromia stringers.

The net mass change curves for AISI 441 exposed at 800, 850 and 900 °C are given in Figs 7.2 to 7.4. Increasing the temperature resulted in an increase of the oxidation rate. In the early stages of exposure the materials behaviour was determined by near-parabolic oxide growth kinetics. The AISI 441 steel oxidised at temperatures in the range 800–900 °C formed at its surface an oxide scale which consisted of an outer $(Mn, Cr)_3O_4$ layer and an inner layer of Cr_2O_3. The internal oxidation zone contained small precipitates of Al_2O_3 and TiO_2.

The net mass change curves for Alloy 800H during exposure at 950, 1000 and 1050 °C are given in Figs 7.5 to 7.7. The early stages of exposure were determined by oxide scale growth showing a near-parabolic time dependence of the scale growth. The time for onset of spallation varied as function of temperature and upper dwell time and in general decreased by decreasing the upper dwell time and for increasing test temperatures. Also the extent of spallation was enhanced by shortening the upper dwell time and/or increasing the oxidation temperatures. During oxidation at 1000 °C, as at 950 °C, the alloy formed at its surface an outer layer of $(Mn, Cr)_3O_4$ spinel and an inner layer of Cr_2O_3, as well as internal oxidation zone. The substantial thermal expansion coefficients mismatch between metal matrix (austenite) and oxide lead however, to spallation of the oxide scale.[1,2] Increasing the oxidation temperature up to 1050 °C resulted in enhanced spallation caused by significant changes of oxide scale morphology and phase composition. The oxide scale formed at 1050 °C was heterogeneous and consisted of Fe, Cr, Ni-spinel, NiO-type oxide and a chromia layer formed near the metal/oxide interface.

Typical mass changes curves for the Ni-based superalloy CM 247 are given in Fig. 7.8. Unlike the aforementioned materials the varied parameters showed only limited (oxidation temperature) or virtually no influence (dwell times) on the oxidation/spallation kinetics of the alloy. The outer part of oxide scales formed on the alloy surface spalled off during the first few cooling cycles. Increasing the oxidation temperature from 1100 to 1150 °C resulted in enhanced spallation during the first 24 h of oxidation.

The net mass change curves for Kanthal A1 are given in Fig. 7.9. All curves show a strong similarity with a strong mass gain during the first cycles and attenuated subsequent oxide growth. The effect of the cycling parameters is very small and does not seem to follow a systematic dependence. Kanthal A1 forms a close scale of Al_2O_3. The test pieces have been subjected to metallographic examination with evaluation of the thickness of the different layers of the test piece cross-sections.

7.2.3 Evaluation of experimental results

For evaluation of the effect of the cycling parameters on the oxidation behaviour, the net mass changes of the test pieces were plotted versus time, as shown in Fig. 7.10a. In the absence of spallation the mass change can be described by Eq. 7.1.

$$\left(\frac{\Delta m_{\mathrm{net}}}{A}\right)^{\mathrm{n}} = k \cdot t \rightarrow \frac{\Delta m_{\mathrm{net}}}{A} = (k \cdot t)^{1/\mathrm{n}} \qquad 7.1$$

From the double logarithmic plot, given in Fig. 7.10b, the oxidation parameters k_n and n can be determined. A change in the mechanism (spallation, breakaway oxidation) becomes apparent by a change of the slope in the double logarithmic plot. Analysis of the linear part of the curve by linear regression by means of simple spreadsheet calculations yields the slope $b = 1/n$ and the y-axis intercept as shown in Fig. 7.10b. The values for the oxidation parameter k_n, the exponent n, the number of cycles $N_{\mathrm{protective}}$ and the accumulated hot dwell time $t_{\mathrm{protective}}$, during which the material shows protective behaviour without spallation or breakaway, are given in Tables 7.2, 7.3 and 7.4 for the materials AISI 441, Alloy 800H and Kanthal A1 respectively. The net mass change data of alloy P91 and CM 247 could not be evaluated by this type of analysis.

Additionally the net mass change difference $\Delta m_{\mathrm{net}}(t_{300\,\mathrm{h}}-t_{\mathrm{protective}})$ as defined in equation 7.2 was determined for all tested materials (Tables 7.2–7.5):

$$\Delta m_{\mathrm{net}}(t_{300\,\mathrm{h}}-t_{\mathrm{protective}}) = \Delta m_{\mathrm{net}}(t_{300\,\mathrm{h}}) - \Delta m_{\mathrm{net}}(t_{\mathrm{protective}}) \qquad 7.2$$

where $\Delta m_{\mathrm{net}}(t_{300\,\mathrm{h}})$ is the net mass change of the test piece after 300 h in mg and $\Delta m_{\mathrm{net}}(t_{\mathrm{protective}})$ is the net mass change of the test piece at the time $t_{\mathrm{protective}}$ which is represented by the last data point in the linear range given in Fig. 7.10b.

7.2.4 Statistical evaluation of experimental results

The test design comprises four factors (temperature, hot dwell time, cold dwell time and humidity), which should be tested at three, three, two and two levels respectively. This results in a test matrix of 36 possible combinations.

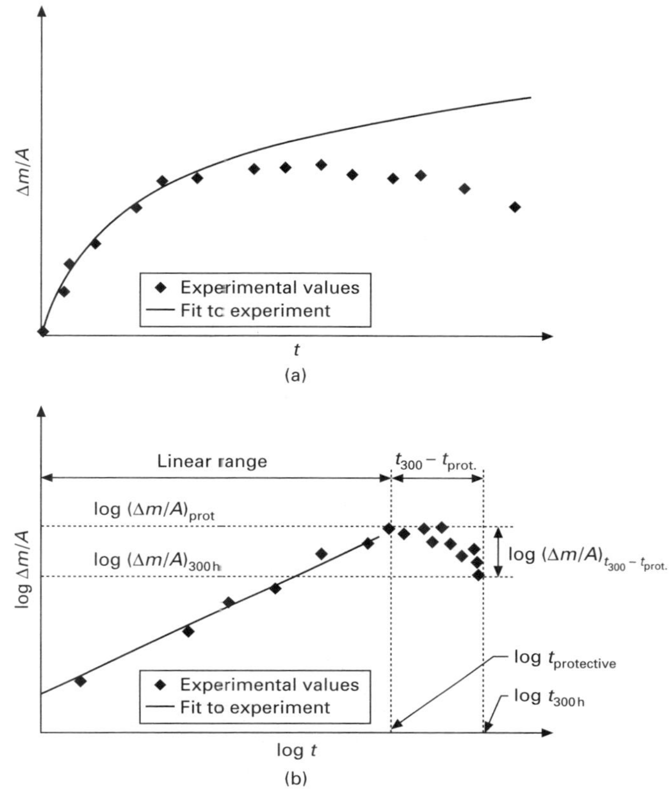

7.10 (a) Plot of net mass change vs time. (b) Double logarithmic plot of net mass change vs time.

To limit the number of experiments a subset of nine experiments was chosen following statistical considerations for a balanced design. The balanced design approach allows investigations of the dependence of the oxidation behaviour on the parameters from a small number of experiments.

There are four outcomes that have been considered in detail. Data are available for the exponent n, and the oxidation parameter k_n, for all of the 18 trials (nine trials for each of AISI 441 and Alloy 800H). Data on the protective oxide growth time $t_{\text{protective}}$ is available for all of the Alloy 800H trials but exceeds the experimental maximum time of 300h for most of the lower temperature trials of AISI 441. The difference of net mass change between $t_{300\text{h}}$ and $t_{\text{protective}}$ is available for all of the Alloy 800H trials but only for the higher temperature AISI441 trials.

The various outcomes are continuous variables and analysis of variance (ANOVA)[3] can be used to determine which parameters have a significant effect on the outcomes. ANOVA assumes that the experimental error is

Table 7.2 Evaluation of experimental net mass change data of AISI 441

	Cycling conditions						
	Dwell time (h)						
T (°C)	Hot	Cold	Environment	n	$k_n \times 10^{-4}$	$t_{protective}$ (h)	$\Delta m_{net}(t_{300h} - t_x)$ (mg)
800	0.5	1	Dry	1.82 ± 0.09	3.76 ± 0.33	>300	
800	1	0.25	Wet	2.21 ± 0.12	0.84 ± 0.23	>300	
800	2	0.25	Dry	1.98 ± 0.20	2.01 ± 0.51	>300	
850	0.5	0.25	Dry	2.06 ± 0.15	8.11 ± 0.28	>300	
850	1	1	Dry	1.42 ± 0.36	8.28 ± 2.06	>300	
850	2	1	Wet	2.20 ± 0.12	3.01 ± 0.47	162	–0.043 ± 0.017
900	0.5	0.25	Wet	2.19 ± 0.11	9.57 ± 1.77	109.5	–0.047 ± 0.014
900	1	0.25	Dry	1.84 ± 0.05	11.67 ± 1.37	174	–0.184 ± 0.043
900	2	0.25	Dry	1.91 ± 0.11	14.77 ± 2.22	120	–0.270 ± 0.024

Table 7.3 Evaluation of experimental net mass change data of Alloy 800H

Cycling conditions							
	Dwell time (h)						
T (°C)	Hot	Cold	Environment	n	$k_n \times 10^{-4}$	$t_{protective}$ (h)	$\Delta m_{net}(t_{300h} - t_x)$ (mg)
950	0.5	1	Dry	2.54 ± 0.09	98.09 ± 3.83	215.5	–5.18 ± 1.80
950	1	0.25	Wet	2.67 ± 0.02	133.22 ± 4.76	133	–1.43 ± 0.05
950	2	0.25	Dry	2.60 ± 0.02	93.81 ± 61	196	–0.23 ± 0.05
1000	0.5	0.25	Dry	2.83 ± 0.05	263.02 ± 125	86.5	–41.92 ± 0.73
1000	1	0.25	Dry	2.71 ± 0.19	258.13 ± 12.87	86	–16.27 ± 2.23
1000	2	1	Wet	2.65 ± 0.06	369.06 ± 15.86	82	–9.90 ± 4.19
1050	0.5	0.25	Wet	2.70 ± 0.08	6390 ± 10.18	46.5	–40.25 ± 0.42
1050	1	1	Dry	2.50 ± 0.05	661.68 ± 19.18	42	–52.91 ± 0.46
1050	2	0.25	Dry	4.06 ± 0.57	301.32 ± 73.39	76	–39.03 ± 0.71

Table 7.4 Evaluation of experimental net mass change data of Kanthal A1

Cycling conditions							
	Dwell time (h)						
T (°C)	Hot	Cold	Environment	n	$k_n \times 10^{-4}$	$t_{protective}$ (h)	$\Delta m_{net}(t_{300h} - t_x)$ (mg)
1200	1	0.25	Lab	2.20 ± 0.10	213 ± 16	> 300	
1250	2	0.25	Lab	2.47 ± 0.29	174 ± 43	> 300	
1250	1	0.25	Lab	2.20 ± 0.28	329 ± 69	> 300	
1200	2	0.25	Lab	2.69 ± 0.14	256 ± 33	> 300	

Table 7.5 Evaluation of experimental net mass change data of heat-treated P 91 and CM 247

	Cycling conditions				
		Dwell time (h)			
Material	T (°C)	Hot	Cold	Environment	$\Delta m_{net}(t_{300\,h})$ (mg)
P91	650	1	0.25	Dry	~ 0
P91	650	1	0.25	Wet	5.62 ± 0.21
P91	650	2	0.25	Dry	~ 0
P91	650	2	0.25	Wet	4.55 ± 0.34
CM 247	1100	1	0.25	Dry	–0.25 ± 0.09
CM 247	1100	2	0.25	Dry	–0.32 ± 0.16
CM 247	1150	1	0.25	Dry	–3.86 ± 0.15
CM 247	1150	2	0.25	Dry	–3.08 ± 0.10

approximately normal, that results are independent of each other and that the variance is constant. A log transformation is applied to the k_n parameter to stabilise the variance of the experimental error.

The experimental design was chosen such that each parameter was tested over the same combination of other parameters, as far as possible. The design gives a balanced comparison of the effects of each parameter. Two parameters were required to be at three levels. This indicated the use of a nine trial design so that each level could be tested three times. The other two parameters therefore occur unequal times for each level. The design was chosen to make the easier level occur more often; therefore there are three wet environments and six dry environments. For the short dwell trials there are three 1 h cold dwell times and six 15 min cold dwell times as this fitted in with laboratory procedures. A reference trial was included in each design.

Statistical evaluation for the example of Alloy 800H

Figure 7.11 shows the main effects for the protective oxide growth time ($t_{protective}$) for Alloy 800H. It can be seen that increasing the temperature has the effect of decreasing $t_{protective}$. Decreasing upper dwell time has the effect of decreasing $t_{protective}$. Environment appears to have virtually no effect. Only the effect of temperature ($p = 0.00$) is statistically significant.

Figure 7.12 shows the main effects for the log of the oxidation rate constant k_n for Alloy 800H. It can be seen that increasing temperature has the effect of increasing log mean k_n. Environment appears to have virtually no effect. Only the effect of temperature ($p = 0.00$) is statistically significant. The effect of dwell times is not large enough to be statistically significant, probably because the nine-trial experiment is not sufficiently powerful.

Figure 7.13 shows the main effects for the difference of net mass change between $t_{300\,h}$ and $t_{protective}$ ($\Delta m_{net}(t_{300\,h}–t_{protective})$) for Alloy 800H. It can be

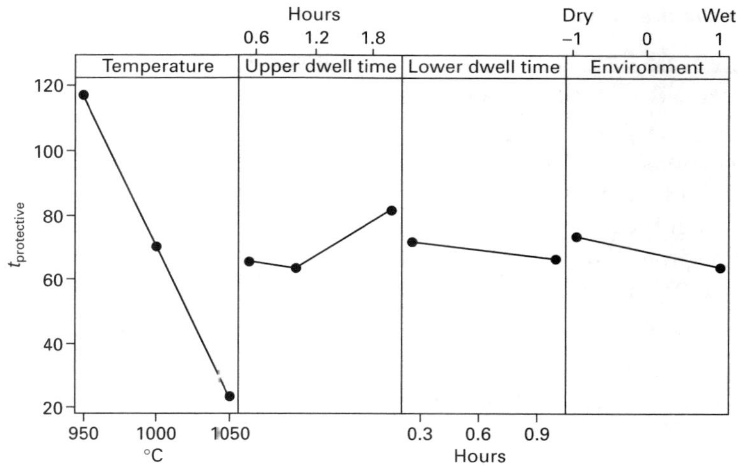

7.11 The main effects plot for the protective oxide growth time ($t_{protective}$) for Alloy 800H.

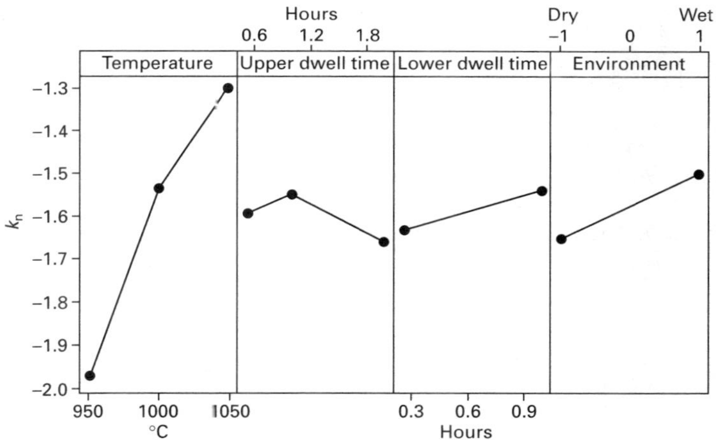

7.12 The main effects plot for the log of the oxidation rate constant k_n for Alloy 800H.

seen that increasing temperature has the effect of increasing $\Delta m_{net}(t_{300h}-t_{protective})$ (in absolute terms). Environment appears to have virtually no effect. Decreasing upper dwell time has the effect of increasing $\Delta m_{net}(t_{300h}-t_{protective})$ (in absolute terms). Only the effects of temperature and upper dwell time are statistically significant. The effect of lower dwell times is not large enough to be statistically significant, probably because the nine-trial experiment is not sufficiently powerful. There were no significant effects on the exponent of the growth law, *n*.

Statistical evaluation for the example of AISI 441

There were insufficient values for a statistical analysis of net mass change, so analysis was carried out for the oxidation rate constant k_n and protective oxide growth time ($t_{protective}$).

The long dwell data show significant effects of upper dwell temperature (Fig. 7.14). In particular, for the log of rate constant, temperature ($p = 0.00$) has a significant effect, with higher temperature increasing the rate constant.

For protective oxide growth time ($t_{protective}$) (Fig. 7.15), the temperature ($p = 0.01$) has a significant effect, with higher temperature decreasing the

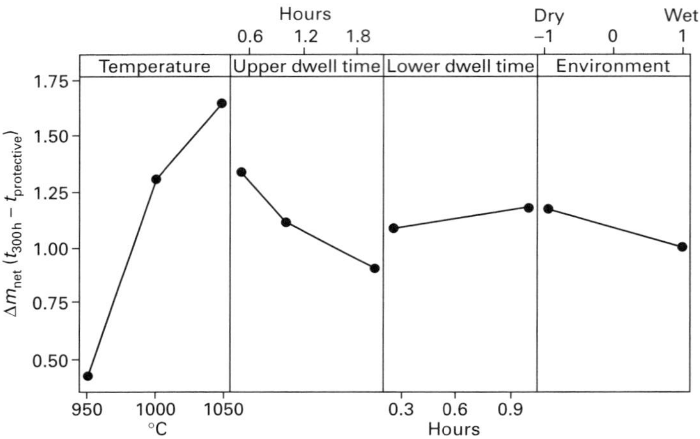

7.13 The main effects plot for the difference of net mass change between $t_{300\,h}$ and $t_{protective}$ ($\Delta m_{net}(t_{300\,h} - t_{protective})$) for Alloy 800H.

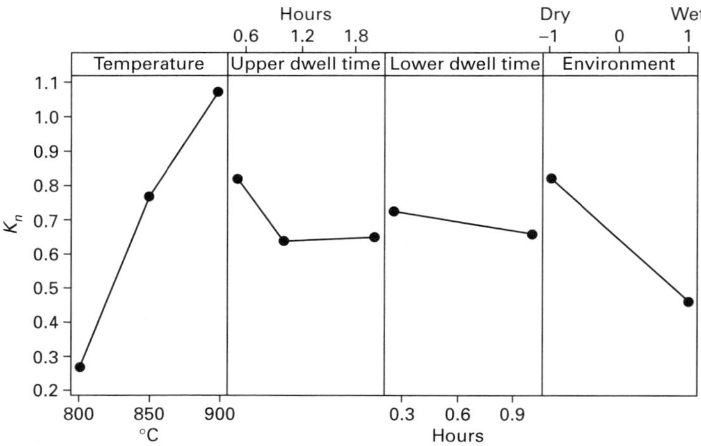

7.14 The main effects plot for log of the oxidation rate constant k_n for alloy AISI 441.

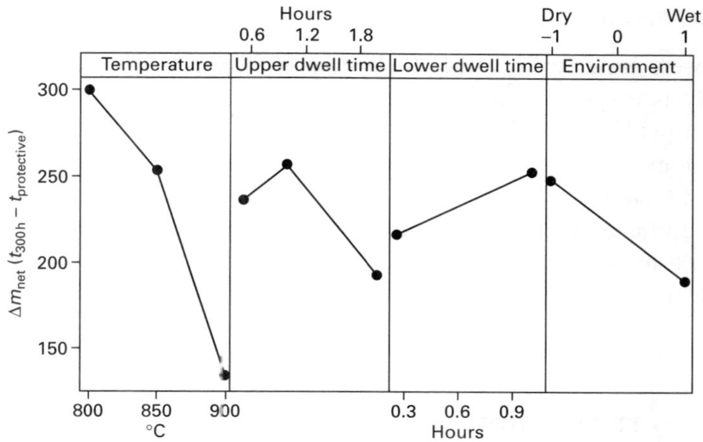

7.15 The main effects plot for the difference of net mass change between $t_{300\,h}$ and $t_{protective}$ ($\Delta m_{net}(t_{300\,h} - t_{protective})$) for alloy AISI 441.

time to spall. Five of the times to spall are greater than 300 h and have been set at 300 h for the analysis.

The data show significant effects of temperature and environment. In particular, for the log of the rate constant, the temperature ($p = 0.00$) and environment ($p = 0.01$) have significant effects, with higher temperature increasing the oxidation rate constant and a wet environment decreasing the oxidation rate constant.

7.3 References

1. L. Antoni, J. M. Herbelin, in *Cyclic Oxidation of High Temperature Materials*, EFC monograph number 27 (Eds. M. Schütze, W. J. Quadakkers), Institute of Materials, London 1999, 187–197.
2. M. A. Harper, B. Gleeson, in *Cyclic Oxidation of High Temperature Materials*, EFC monograph number 27 (Eds. M. Schütze, W. J. Quadakkers), Institute of Materials, London 1999, 273–286.
3. J. R. Nicholls, P. Hancock (1983) 'The analysis of oxidation and hot corrosion data – a statistical approach', in *High Temperature Corrosion*, (Ed. R. A. Rapp) NACE-6 Houston, pp 198–210.

8

Investigation of the influence of parameter variation in ultra-short dwell thermal cycling oxidation

J. R. NICHOLLS and T. ROSE, Cranfield University, UK

8.1 Introduction

Ultra-short dwell time tests find wide application in industry, a major market being electrically heated heating elements that are in use in many furnace installations that are used across a wide sector of the manufacturing industry, including alloy manufacture, food processing, commercial heat treatment, coating manufacture, pottery, automotive and aerospace industries. Other applications for heating elements in the domestic market include space heaters, water heaters and many other domestic appliances. Applications not utilising Joule heating are also of importance, particularly foil-based automotive catalyst support carriers, burners, hot gas filters, rocket engine nozzles, etc., and these applications have an important role to play in the power generation, chemical, automotive and aerospace industries. Such components are manufactured from wire, foil or strip product, but not necessarily so. Ceramic materials are also widely used in applications involving ultra-short cycle testing. Thus a major objective of the 'ultra-short test standardisation' was to define a test method with wide applicability to all industrial sectors.

8.2 Definition of suitable test conditions

From discussion within this partnership the following test conditions were considered most appropriate for a standard 'ultra-short dwell', rapid thermal cycle test procedure:

- Maximum temperature – 1300 °C.
- Heating period – 2, 5, 10 min.
- Cooling period – 2, 5 min.
- Maximum cycle count – 1200 cycles, or failure if earlier.
- Maximum time at temperature – 100 h.

Secondary parameters, which will influence performance but are not easily controllable, include:

- heating rate;
- dwell time at maximum temperature;
- cooling rate.

These secondary, or deduced, parameters have a direct influence on the extent of oxidation; the morphology, microstructure and phase content of the oxide; the stress generated within the system on heating/cooling; the rate of scale-forming element consumption and hence the depletion in the available elemental reservoir and hence on component life. These observations apply whether the component is metallic, intermetallic, ceramic, cermet or composite. Ideally, the chosen method should be able to test all these forms of component. At present, the largest market sector for this form of testing evaluates wire and foil products.

8.3 Possible alternative test procedures

Three alternative test procedures were evaluated as processes to provide a standard test method. These procedures were Joule heating, induction heating and focused light (including laser heating methods).

8.3.1 Joule heating

This method is widely used by all resistance heating wire manufacturers and can be applied to foil components with only minor modifications to existing test systems. For this sector of the industry (resistance heating element manufacture) this is in effect the de facto standard, although details of the precise methods used may vary between manufacturers. Thus, Joule heating provides a simple method, aligned to procedures used by one sector of the industry. It can be used for foil samples with some limited modification but cannot be easily applied to other product forms or more complex component geometries.

Major limitations, as far as wide applicability, are that the sample must be electrically conducting – precluding ceramics, many cermets and some high-resistance metallic and intermetallic materials – and the problems of component geometry mentioned above. The sample heating rate and cooling rate depend on sample geometry and material heat capacity. If hot spots are formed – a result of local section loss – local temperatures may rise, giving early failure. Thus precise control of maximum temperatures and time at temperatures is difficult. It is not possible to directly measure mass gains and thus assessment methods are not directly comparable with methods used for the 'long dwell' and 'short dwell' test methods. Pyrometric methods and changes in component resistance can be used to measure temperature.

8.3.2 Induction heating

This method is less widely used, but can be used for 'contact-free' assessment of the oxidation of both solid and liquid metallic alloys. Again the method is limited to electrically conducting (or semiconducting) samples. Geometry constraints are more relaxed, thus many conventionally shape sample geometries – including block, disc and cylinders – can be studied. The temperature cycling rate depends on the sample heat capacity and radiation and convection effects within the test cell. Pyrometric methods have to be used to measure temperature.

8.3.3 Focused light and laser heating

Focused light and laser heating methods offer a method of direct heating with a rapid switchable power source. Large solar furnaces exist in Spain and France, and lasers have been widely used for short duration surface heat treatment and allow power densities sufficient to melt and weld components.

There are no major material limits, provided the materials are capable of high-temperature operation. However, the efficiency of heating depends on the local surface emissivity. Problems of high reflectance surfaces can be overcome by modified surface finishes. Surface oxidation is a benefit as it generally improves the emissivity of a surface. There are no geometrical constraints and conventional thermogravimetric methods could be used to monitor oxidation mass gains. Thus block, disc and cylinder specimens could be used – directly comparable with data gathered in the 'long dwell' and 'short dwell' test programmes.

Measurements of temperature would involve embedded thermocouples, or pyrometry. Temperature cycling depends on the sample heat capacity, emissivity, radiation and convection interactions with the test equipment.

8.4 Design of a 'focused light' rapid thermal cycle test facility

Conceptually, the design of a dual, focused light, thermobalance facility is illustrated in Fig. 8.1. The aim was to rapidly cycle (ultra-short dwell) one specimen, while compensating for buoyancy effects using a second (dummy) sample that will not oxidise. Following discussion with COTEST partners, it became apparent that an optimised system should have the capability of testing multiple specimens, along with the ability to measure both mass gain and resistance change during the rapid thermal cycle process.

To achieve these aims a new 'focused light', rapid thermal cycle furnace has been designed. A cross-section of the design is illustrated simply and schematically in Fig. 8.2. It is designed to be operated in a vertical mode,

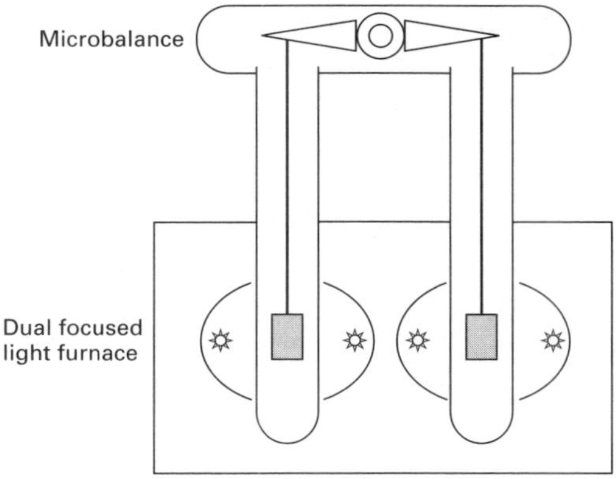

8.1 Focused light microbalance.

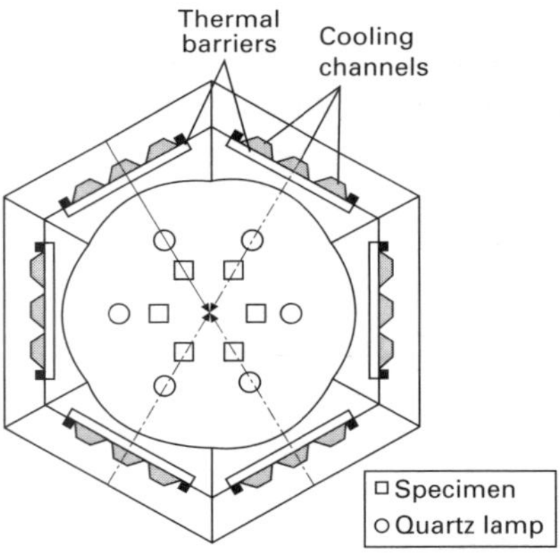

8.2 Focused light, multi-sample furnace.

using six 500 W (3 kW total) or six 1000 W (6 kW total) quartz lamps depending on the specimen temperature that must be achieved. The furnace utilises six parabolic mirror segments to provide the overall hexagonal design. Each mirror segment consists of two stainless steel parts, separated by a thermal barrier and water cooled using a labyrinth cooling channel. The two parts are 'O' ring sealed and the parabolic mirror surface is highly polished and plated with gold, to provide low emissivity (ε_{tot} = 0.02–0.03) for metal surface temperature in the range 100–600 °C. In the near infrared (λ = 0.65 μm) the spectral emissivity of gold (0.14) is second only to silver (0.07); silver is, however, susceptible to tarnishing. Thus some 2–3% of the incident energy would be absorbed, providing the reflecting surfaces are kept relatively cool (below 600 °C). For a 1 kW segment load, the reflector surface temperature is expected not to exceed 300 °C.

The specimen positions are heated by direct radiation from the nearest adjacent quartz lamp and 'focused light' from the diametrically opposite quartz lamp. Thus three pairs of identically heated specimens are provided in this design. Specimens can be either wire, foil, strip, block, disc or rod shaped and can be monitored by any of the accepted methods: weight change, resistance change or continuous thermogravimetry.

Although this conceptual design of a 'focused light' furnace has been completed, no prototype has been built, as after 6 months it was decided that the ultra-short cycle tests should focus on Joule heating, as this was the de facto standard used in the heating element industry. Design work into a Joule heating system has continued.

8.5 Design of a Joule heating device for wire and foil materials

The design of the Joule heating system follows much more traditional lines, but is fully microprocessor controlled. The design is illustrated schematically in Fig. 8.3, with the control loop for a single station reproduced in Fig. 8.4. By using the 'U' shaped specimen design (length of free wire segment 230 mm) then the catenary is fully developed and any further expansion will

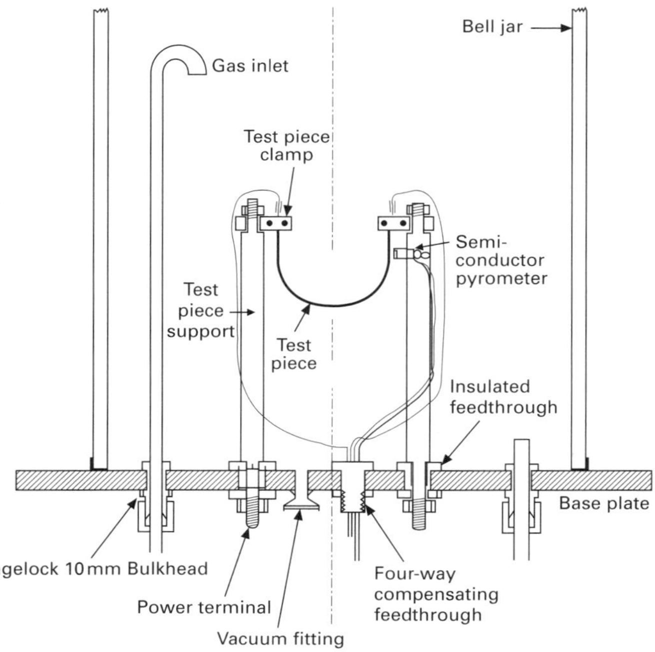

8.3 Joule heating rapid thermal cycle rig.

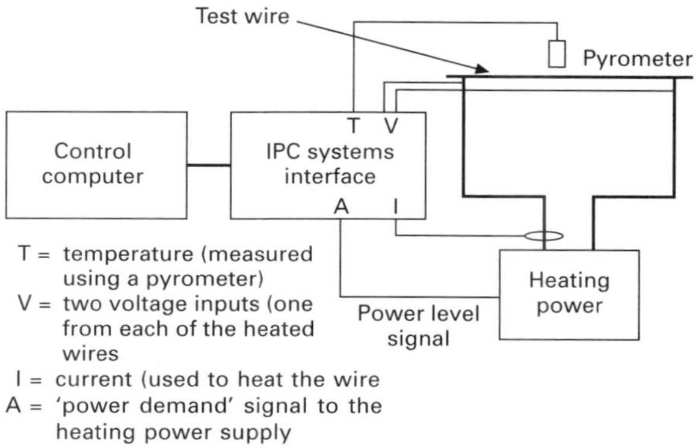

8.4 Control schematic for Joule heating rig.

not significantly alter the distribution of bending stress seen by the wire loop. Equally this approach leads to a compact design and ensures the pyrometric sensor always maintains the same geometry relative to the thermal cycling wire. Such a design allows a much more simple, bell-shaped environmental cell to be constructed with evacuation, environmental control and environment monitoring mounted into the base plate. A prototype test cell was constructed to evaluate the feasibility of this approach.

The microprocessor controlled (IPC) control loop is illustrated schematically in Fig. 8.4. The processor monitors emitted radiation (a function of temperature) through the use of modern semiconductor pyrometric sensing elements, the voltage drop generated across the wire, the current flowing in the wire and based on these three signals provides a power level signal to drive the heating electronics. This control circuitry is well proven in 'weld simulator' studies at Cranfield, which have similar monitoring and control needs. A master computer provides the demand signal and has the capability of monitoring and controlling up to eight systems with any pre-programmable wave form. The proposed design will permit current, voltage and emitted energy monitoring, plus the control of heating power on any time-dependent function of these three parameters. Thus it will be possible to control the system in 'voltage mode', 'current mode' or 'power mode' and pre-programme heating waveforms based on pyrometer reading (a function of temperature, which must be calibrated) resistance, current, voltage or power. Additional functionality will include the control of environment, evacuation and pump-down sequences, plus monitor of the test chamber pressure, water vapour content and possibly oxygen partial pressure if required.

Figure 8.5 illustrates the temperature calibration for Joule heating of Kanthal A1 wire and strip while Figs 8.6–8.8 illustrate typical ultra-short cycle tests on Kanthal A1 at 1200, 1250 and 1300 °C using various test geometries. The new facility was modified from the ASTM standards to permit tests in controlled

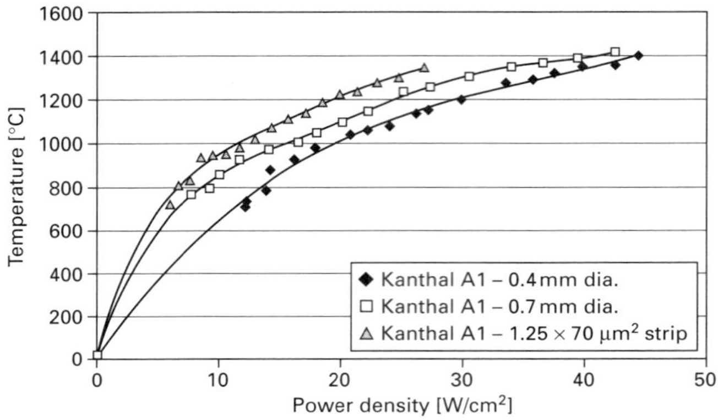

8.5 Temperature calibration for Joule heating of Kanthal A1 wire/strip.

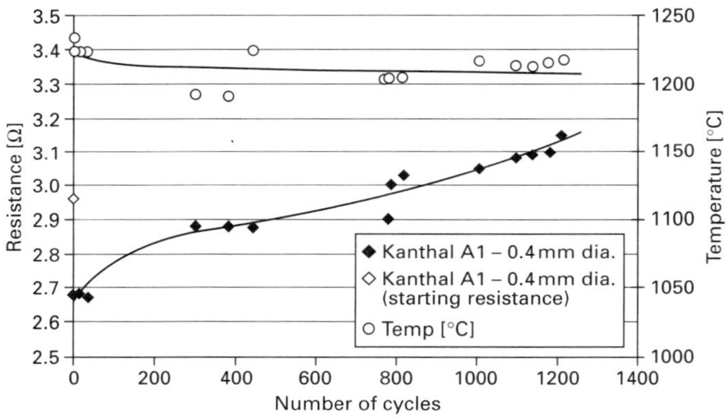

8.6 Ultra-short cycle test of Kanthal A1 at 1200 °C, 2 min hot/2 min cool in air, 0.4 mm diameter wire.

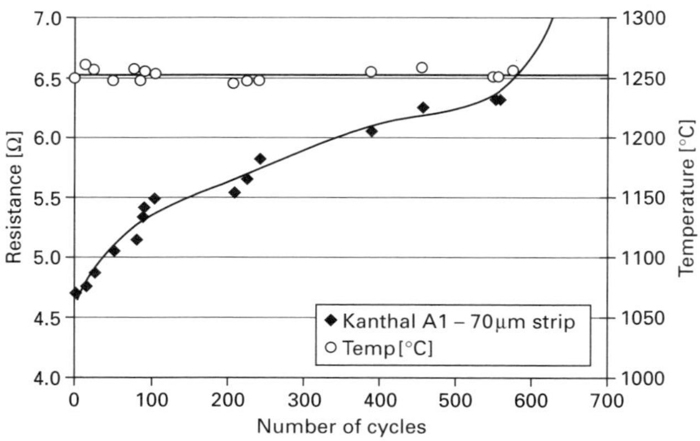

8.7 Ultra-short cycle test of Kanthal A1 at 1250 °C, 5 min hot/2 min cool in air, 70 μm thick ribbon.

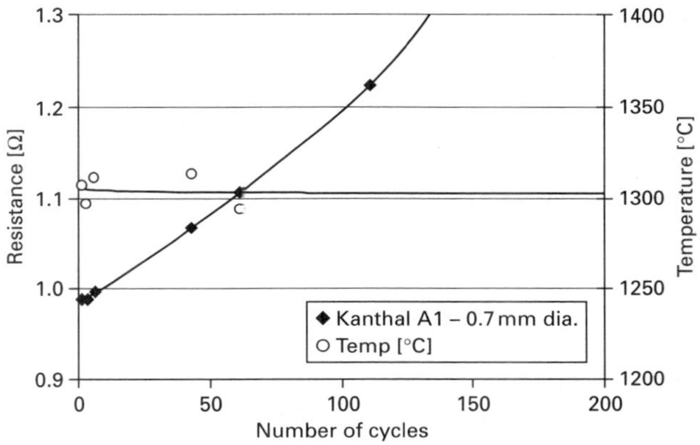

8.8 Ultra-short dwell cyclic oxidation of Kanthal A1 at 1300 °C, 5 min hot/2 min cool in air, 0.7 mm diameter wire.

environments, thus allowing for the first time the rapid thermal cycle testing of wire and ribbon samples in controlled atmospheres.

8.6 Ultra-short dwell experiments

The 'ultra-short dwell' test matrix is aimed at maximising the available test resource (four machines) and testing time, while evaluating available materials. The following materials have been tested:

- Alloy 800; 0.4 mm diameter wire;
- Kanthal A1; 0.4 mm diameter wire;
- Kanthal A1; 0.7 mm diameter wire;
- Kanthal A1; 70 µm × 1.25 mm wide foil.

Thus testing the Kanthal A1 materials allows the role of specimen geometry to be evaluated, while tests comparing Alloy 800 and Kanthal A1 are possible for 0.4 mm diameter wire. Three independent sets of nine tests, using a full 3 × 3 test matrix and including triplicate test specimens have been undertaken and all tests in air and two levels of water vapour (4 vol% and 10 vol%) have been completed.

8.6.1 Role of specimen geometry

Tests were undertaken using Kanthal A1 in laboratory air, with three upper dwell temperatures (e.g. 1200, 1250 and 1300 °C) and three upper dwell times (2, 5 and 10 min). The sample geometries evaluated were 0.4 mm wire, 0.7 mm wire and 70 µm × 1.25 mm foil. The lower dwell time was kept constant at 2 min (originally it had been planned to use a 5 min lower dwell time however, experimental studies had shown that it cools to room temperature well within 2 min, thus a 2 min lower dwell allowed to shorten the overall test programme without impacting the outcomes of the test matrix).

Originally it had been planned to complete duplicate tests, but to increase the reliability of the data, with four rigs available, triplicate samples were tested. Table 8.1 summarises the measured cycle lifetimes and change in resistance (%) at failure, on test completion. The change in resistance in the early part of the test cycle can be directly related to the section loss due to cyclic oxidation.

Figures 8.6, 8.7 and 8.8 illustrate three typical sets of results for the three geometries. Figure 8.6 illustrates the change in resistance and temperature (the test is controlled on power applied to the wire) through a 2 min upper dwell, 2 min lower dwell test cycle. Under this test condition the Kanthal A1 0.4 mm diameter wire survives 1200 test cycles. There is an initial resistance change on first heating to temperature, followed by a gradual increase with the number of cycles from 2.68 to 3.15 Ω. The temperature varied between 1191 and 1224 °C, with the first cycle marginally hotter at 1233 °C. Thus the mean test temperature was 1212 °C.

Figures 8.7 and 8.8 show how raising the test temperature to 1250 and 1300 °C rapidly reduces the wire cyclic lifetime. In each case the wire resistance increase throughout the lifetime, often changing by up to 40% before failure. The most damaging cycle was the 2 min hot/2 min cool cycle at any given temperature.

8.6.2 Role of environment

These tests have been undertaken using the 0.4 mm diameter Kanthal A1 and the 0.4 mm diameter Alloy 800 wires. The upper dwell temperature has to be different for the two alloys (e.g. 1200, 1250 and 1300 °C for Kanthal A1 and 950, 1000 and 1050 °C for Alloy 800) but the environments examined were the same, otherwise the Alloy 800 has too short a life. Again for the Kanthal A1 a full 3 × 3 matrix was completed, thus three upper dwell temperatures × three upper dwell times × three environments have been examined. Table 8.2 presents the test matrix for Kanthal A1.

With triplicate specimens and an average test duration of 100 h, this involved over 4000 h of testing. Had all specimens lasted 1200 cycles the total test time would have been 5400 h. Many of the samples when tested at 1300 °C failed earlier than 1200 cycles; also at 1250 °C, a few samples fail at under 100 h (1200 cycles) of testing.

The three environmental conditions examined were dry air, 4% water vapour or 10% water vapour. Figures 8.9–8.11 summarise the change in resistance measured for Kanthal A1 in these environmental tests. Again each test condition was repeated three times. As is evident from Figs 8.9–8.11,

Table 8.1 Test matrix for Kanthal A1 – specimen geometry and summary of results

Trial	Upper dwell temp. (°C)	Upper dwell time (min)	Specimen geometry	Cycle lifetime*	Change in resistance (%)
1	1200	2	0.4 mm ϕ wire	952, 1206, 1273	4.55, 5.11, 13.65
2	1200	5	0.7 mm ϕ wire	3 × >1200	5.62, 31.53, 33.73
3	1200	10	70 μm foil	805, 815, 1004	33.06, 43.03, 49.41
4	1250	2	0.7 mm ϕ wire	3 × >1200	17.13, 23.73, 37.12
5	1250	5	70 μm foil	90, 611, 638	3.86, 39.52, 39.63
6	1250	10	0.4 mm ϕ wire	39, 1200, 1207	0.00, 21.83, 22.02
7	1300	2	70 μm foil	521, 673, 1153	43.22, 36.52, 34.53
8	1300	5	0.4 mm ϕ wire	115, 1112, 1204	19.10, 37.04, 47.44
9	1300	10	0.7 mm ϕ wire	612, 641, 677	31.14, 34.36, 41.47

* Number of cycles; tests were terminated if the wire survived 1200+ cycles.

Table 8.2 Test matrix for Kanthal A1 – environmental effects

Trial	Upper dwell temp. (°C)	Upper dwell time (min)	Environment
1	1200	2	Dry air
2	1200	5	4% specific humidity
3	1200	10	10% specific humidity
4	1250	2	4% specific humidity
5	1250	5	10% specific humidity
6	1250	10	Dry air
7	1300	2	10% specific humidity
8	1300	5	Dry air
9	1300	10	4% specific humidity

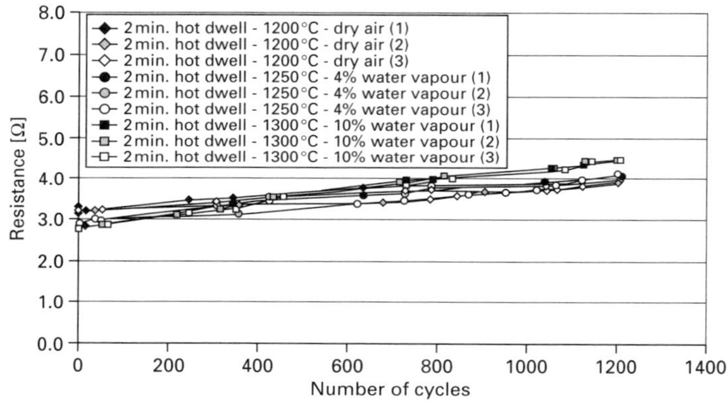

8.9 Ultra-short dwell tests on Kanthal A1 wire (0.4 mm diameter) for 2 min hot dwell per cycle in various environments.

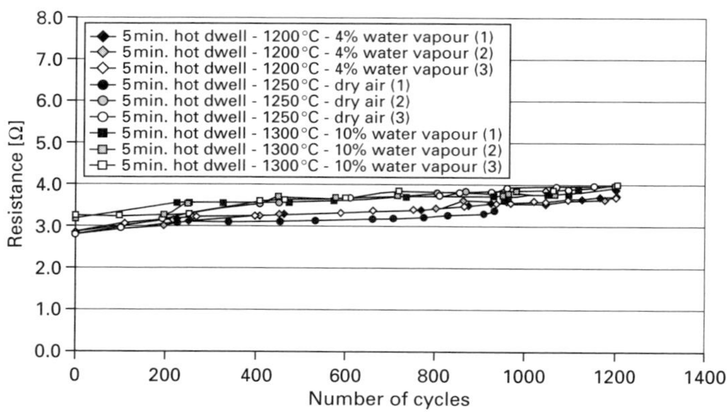

8.10 Ultra-short dwell tests on Kanthal A1 wire (0.4 mm diameter) for 5 min hot dwell per cycle in various environments.

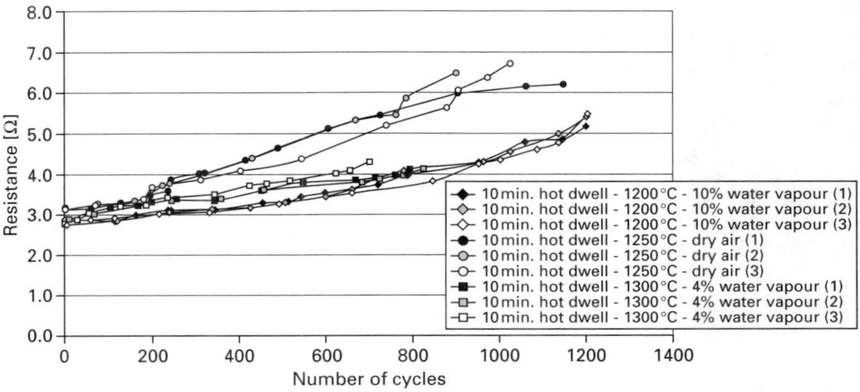

8.11 Ultra-short dwell tests on Kanthal A1 wire (0.4 mm diameter) for 10 min hot dwell per cycle in various environments.

repeatability between tests for given test conditions and environment is excellent.

At 1200 °C all samples survived 1200 cycles, whether tested in dry air, 4% or 10% water vapour and for hot dwell times of 2, 5 or 10 min (1200 cycles × 10 min hot dwell time is a 200 h rapid cyclic oxidation test). At 1250 °C, both the 2 and 5 min hot dwell specimens survived 1200 cycles. These were both tested under humid air conditions. However, in the 1250 °C test of 10 min hot dwell in dry air only one sample reached 1200 cycles without failing.

At 1300 °C, the observations are similar to those at 1250 °C, in that the 10 min hot dwell specimens failed before 1200 cycles. The environment under this test condition was 4% water vapour. Thus from these rapid cyclic oxidation tests of Kanthal A1 in humid environments one can conclude that:

- the wire lifetime decreases with increase in temperature;
- as observed from the geometry studies, the 2 min hot dwell/2 min cold dwell is the most severe of test conditions;
- water vapour influences the cyclic lifetime (see Fig. 8.12). The minimum lifetime for Kanthal A1 is observed for tests with a 4% water vapour content.

Figure 8.13 illustrates the cyclic oxidation of Alloy 800H using a 0.4 mm diameter wire in dry air at 950 °C, using a 2 min hot/2 min cool test cycle. No failure was observed in the 1200 cycle test duration. The trend observed was similar to that for Kanthal A1 at 1200 °C: first an initial drop in resistance, followed by a gradual increase from 2.45 to 2.65 Ω (some 6–7% change in measured resistance) throughout the duration of the test, where the gradual increase in resistance with number of cycles is due to the loss in wire section as a result of the cyclic oxidation process. Finally, a more rapid resistance

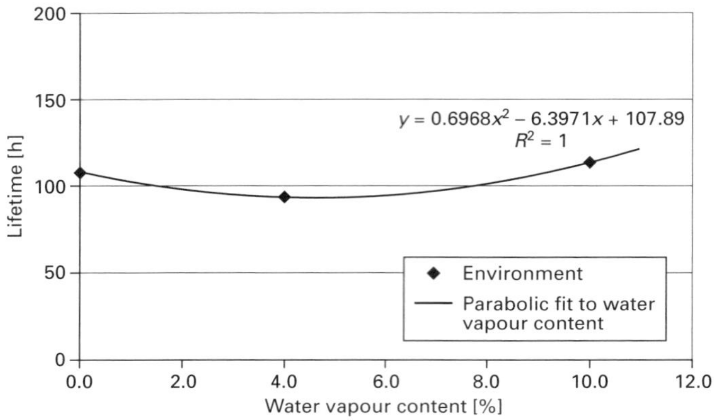

8.12 Influence of water vapour content on the ultra-short cyclic oxidation behaviour of Kanthal A1.

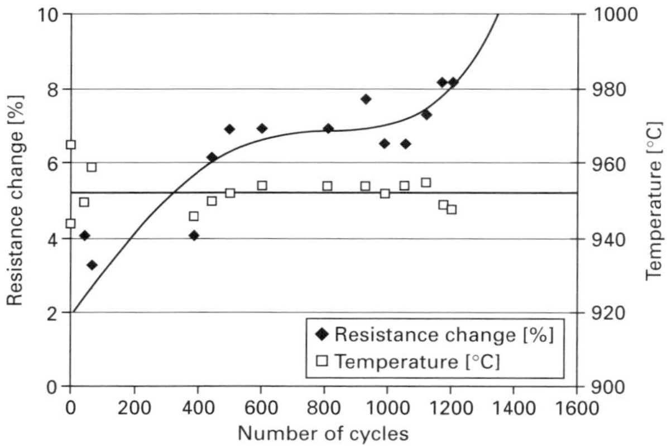

8.13 Alloy 800–950 °C, 2 min/2 min in air.

increase starts as the end of life is approached. However, this wire did not fail at the end of this 1200 cycle test.

When tested at 1000 °C, the lifetime of Alloy 800 varied between 530 and 587 cycles. These tests were undertaken using the 5 min hot, 2 min cool cycle – part of the validation test matrix. However, in terms of accumulated lifetime, measured in hot hours, 500+ cycles of 5 min gives a lifetime of 42+ h, just larger than the lifetime at which the 950 °C test was terminated, without failure.

When tested under the environmental test conditions (4 vol% water vapour and 10 vol% water vapour in air) using the 5 min hot/2 min cold cycle mean cycles to failure are further reduced, and the scatter in the lifetimes increased. This can be seen in Table 8.3, which compares the 1000 °C/5 min hot/2 min cold ultra-short thermal cycle data as a function of the environment (tests undertaken to validate the ultra-short cycle test procedure).

Table 8.3 Ultra-short cycle test on 0.4 mm diameter Alloy 800 at 1000 °C using a 5 min hot/2 min cold test cycle

Environment	Mean test temperature (°C)	Cycle lifetime	Resistance change (%)
Dry air	999	530, 563, 587	0.0, 4.85, 5.19
Air–4 vol% H_2O	999	413, 512, 667	4.23, 6.87, 6.98
Air–10 vol% H_2O	1001	392, 433, 638	8.35, 15.39, 17.80

8.6.3 Analysis of resistance change for Joule heating of wire samples

To evaluate the effect of cyclic parameters on oxidation behaviour, it is common practice to plot net mass gain of the test piece versus time. For studies using ultra-short dwell rapid cyclic oxidation by Joule heating this mass change approach to monitoring oxidation is not possible because any demounting of the specimens from the test fixtures would result in a wire breakage.

It is proposed to use hot resistance change as an alternative, but parallel, oxidation monitoring tool. The concept relies on cyclic oxidation producing a reduction in metal cross-section. This cross-section then leads to a progressive increase in resistance that can be measured by monitoring the hot current and associated voltage drop across the wire under test. The fractional change in resistance, where R_o is the minimum measured hot resistance early in the test cycle, is given by:

$$\frac{\Delta R}{R_o} = \frac{R(t_n) - R_o}{R_o}$$

where $R(t_n)$ is the measured hot resistance at time t_n, during the nth cycle. $R(t_n)$ and R_o are measured in ohms.

Plotting the change in resistance against number of cycles results in a curve similar to that plotted in Fig. 8.13 for Alloy 800H or Fig. 8.14 for Kanthal A1. Since the resistance of the wire can be calculated from its change in dimensions provided the resistivity is known, then the influence of dimension change due to oxidation can be related to observed changes in wire resistance.

$$\frac{\Delta R}{R_o} = \frac{\rho\dfrac{L(t_n)}{A(t_n)} - \rho\dfrac{L_o}{A_o}}{\rho\dfrac{L_o}{A_o}}$$

where ρ is the resistivity of the wire, L_o and A_o the original wire length and cross-section and $L(t_n)$ and $A(t_n)$ the wire dimensions after an exposure time t_n.

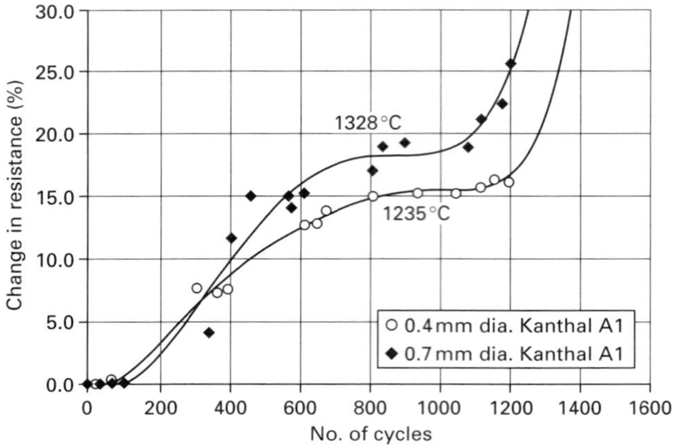

8.14 Change in resistance with number of cycles.

The full analysis is presented in the guidelines. For small changes in wire dimensions, a simplified version of the above equation permits an estimate of metal loss (Δx) due to oxidation

$$\Delta x \simeq \frac{d_0}{4} \cdot \frac{\Delta R}{R_0}$$

where d_o is the original diameter of the wire.

Table 8.4 summarises the calculated metal loss and measured lifetimes for the environmental tests on the Kanthal A1 specimens. It can be seen from Table 8.4 that as a general observation metal loss increases with exposure time. This is observed at 1200 and 1250 °C, irrespective of environment. However, at 1300 °C the shortest lifetime is found for the 10% water environment but this gives the maximum metal loss, as determined by change in resistance. Therefore for this alloy, Kanthal A1, the effect of water vapour would appear more severe at the higher test temperature.

8.7 Conclusions

- A test procedure for testing wire and foil samples under different environmental conditions has been developed, using rapid thermal cycling by Joule heating.
- The methodology permits ultra-short dwell high-temperature exposure, 2–10 min, with rapid cooling rates – from 1300 °C to room temperature in less than 2 min.
- The test procedure meets, and can test to ASTM 76B and 78B, while extending these test standards by permitting wire and foil samples to be evaluated in controlled gaseous atmospheres.

Table 8.4 Calculated metal loss for Kanthal A1 following environmental tests

Temperature (°C)	Hot dwell time (min)	Water vapour (%)	Change in resistance (%)	Mean metal loss (μm)	Lifetime (h)
1200	2.0	0.0	20.27, 23.86, 24.07	22.7	40.15
1200	5.0	4.0	30.79, 33.82, 30.62	31.7	100.20
1200	10.0	10.0	8293, 86.91, 91.35	81	200.44
1250	2.0	4.0	43.58, 40.78, 43.74	42.7	40.26
1250	5.0	10.0	40.70, 44.05, 44.56	43.1	100.25
1250	10.0	0.0	919, 108.62, 115.68	102	183.28
1300	2.0	10.0	62.53, 53.06, 63.58	63.06	40.10
1300	5.0	0.0	25.17, 26.44, 24.99	25.5	100.22
1300	10.0	4.0	466, 48.71, 54.83	50.4	139.89

- The test procedure has been evaluated statistically, using a Taguchi designed balanced subset of the 27 tests in a 3 × 3 × 3 test matrix. Two sets of nine conditions were evaluated – specimen geometry and environmental effects – with three repeat samples at each test condition
- An analysis of variance (ANOVA) of this ultra-short dwell test series confirmed that temperature, hot dwell time, sample geometry and environment (in this test series, water vapour) were significant factors controlling the cyclic lifetime of wire or foil samples. A notable interaction between these main factors was also observed.
- When the hot endurance (wire/foil hot lifetime) is calculated by multiplying the 'hot dwell time' by the number of cycles, then hot dwell time becomes the major life limiting factor, with short duration 'hot dwells' (2 min on/ 2 min off) cycle the most severe.

9 Burner rig thermal cycling oxidation testing

A. KLIEWE, MTU Aero Engines GmbH, Germany and
S. OSGERBY, NPL Ltd, UK

9.1 Introduction

Burner rig testing replicates the simultaneous combustion and deposition processes that occur during operation of certain types of high-temperature plant, e.g. gas turbines. Two types of burner rig are available, differentiated by the velocity of the combustion atmosphere through the system: low-velocity burner rig, with a hot gas velocity up to 100 m/s and high-velocity burner rigs with more than 100 m/s hot gas velocity.

The test involves injection of fuel (e.g. kerosene, oil or gas) and salt solution to simulate impurities which are typical for the particular application into a heated chamber where combustion occurs. The combustion atmosphere then carries the salt downstream where it may react with the fuel and/or vaporise; it is subsequently deposited/condensed on the test pieces. The combustion and stabilisation zones of the burner rig are maintained at a higher temperature than the test piece zone in order to provide the driving force for deposition. The test pieces may be cooled (high-velocity rigs) to simulate the heat flux experienced in the turbine, or rotated (low-velocity rigs), to ensure homogeneous deposition of salt around their circumference.

9.2 Low-velocity burner rig

The critical parameters for low-velocity burner rig testing were previously identified[1, 2] and include (i) salt deposition rate and (ii) combustion fuel. Periodic removal of deposits was also identified as a potentially critical parameter during initial discussions between the COTEST partners. Flexibility in other test parameters (e.g. cycle times) is restricted due to the manual nature of this type of testing. The testing schedule was agreed and is detailed below:

- Three materials (P91 at 650 °C, AISI 441 at 850 °C and Alloy 800H at 950 or 1000 °C).

- 24 h cycles (20 h at high temperature, 4 h at ambient temperature).
- Continuous salt deposition at two rates ('normal' and 'half normal').
- Deposit removal every four cycles on 50% of specimens.
- Specimens removed after 4, 8 and 16 cycles.
- Mass change measurements after 1, 2, 3, 4, 6, 8, 12 and 16 cycles.
- Deposit: K_2SO_4.
- Fuel: low-S kerosene, kerosene + 1% S.
- Visual examination and digital image recorded after each cycle.

In order to synchronise testing with the working day, the thermal cycle was fixed at 24 h (20 h on test with 4 h available for cooling and measurement). Test pieces were removed from testing at the end of each working week and stored in desiccators for the weekend. Mass change and macro-appearance were recorded. Some pins (numbers 1, 3 and 5) were washed by agitation in de-ionised water at the end of every four cycles. After the start of the test programme it was decided that it was preferable to use salt composition as a test variable rather than deposition rate and a second salt ($Na_2SO_4 : K_2SO_4$) included (Table 9.1). Inert pins, manufactured from recrystallised alumina, were included in each temperature cycle. The mass gain of these pins gave a measure of the amount of salt that was deposited during each test.

Within each individual run the mass change data were reproducible between individual pins until spallation occurred. After this point differences in mass change were observed between pins that were washed periodically and those that were not. An example of this behaviour is shown in Fig. 9.1.

Table 9.1 Test matrix for low-velocity burner rig testing

Fuel	Salt deposit	Contaminant flux rate, (mg/(cm^2 h))	Max. number of cycles	Specimen handling
Kerosene	K_2SO_4	2.33	16	Deposit removed by washing after every four cycles Deposit allowed to build up continuously
Kerosene + 1% S	Na_2SO_4/K_2SO_4	2.33	16	Deposit removed by washing after every four cycles Deposit allowed to build up continuously
	Na_2SO_4/K_2SO_4	2.33	16	Deposit removed by washing after every four cycles Deposit allowed to build up continuously

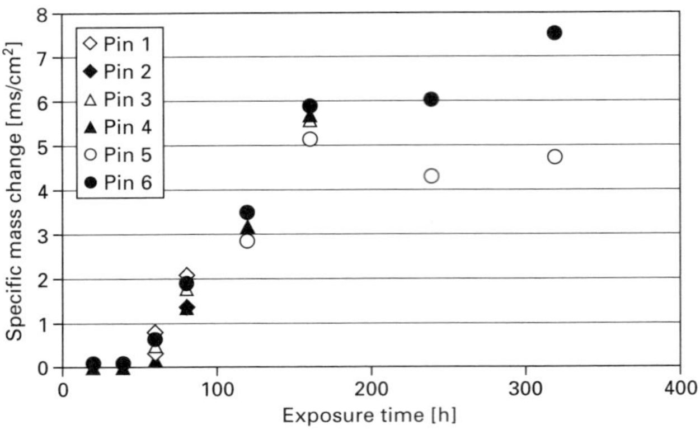

9.1 Mass change data for P91 exposed in burner rig to K_2SO_4 salt.

Mass change data for those pins exposed for the full duration of the test are illustrated in Figs 9.2–9.4 for the three materials. Figure 9.2 shows that for P91 the initial rate of mass increase was greater when using the K_2SO_4 salt than the Na_2SO_4/K_2SO_4 mixture. Using the K_2SO_4 salt, spallation occurred after eight cycles, i.e. when the specific mass change reached ~6 mg/cm^2. Spallation was not observed when the Na_2SO_4/K_2SO_4 salt mixture was used, although the critical specific mass change of ~6 mg/cm^2 was not reached until the end of the test.

Figure 9.3 shows the specific mass change data for AISI 441, exposed at 850 °C. Spallation occurs immediately following the first cycle for both salt compositions. Towards the end of the test, breakaway corrosion, evidenced by rapid mass change, is observed in the test using K_2SO_4 salt. The scatter between individual test pieces was higher in this series of tests than for P91. This is attributed to test piece geometry – AISI was available only in thin sheet form; test pieces were therefore rectangular in shape and accelerated attack at the corners was observed. This geometry of test piece is not recommended if it can possibly be avoided.

Figure 9.4 shows the specific mass change data for Alloy 800H, exposed at 850 °C. Again spallation occurs from the first cycle of the test under both salt compositions. After six cycles the test using K_2SO_4 salt begins to show a mass increase although spallation of the corrosion product is still occurring, as evidenced by the influence of washing the pins. The test under Na_2SO_4/K_2SO_4 salt mixture continued with steady mass loss due to spallation.

Figure 9.5 shows the influence of fuel on net mass change during exposure. For both P91 at 650 °C and Alloy 800H at 950 °C the increased sulphur in the fuel increases the net specific mass change.

Pins were sectioned and examined metallographically after exposure. There is an apparent discrepancy between the mass gain values in Fig. 9.2 and the observed scale thickness. A corrosion product is formed on AISI 441 after

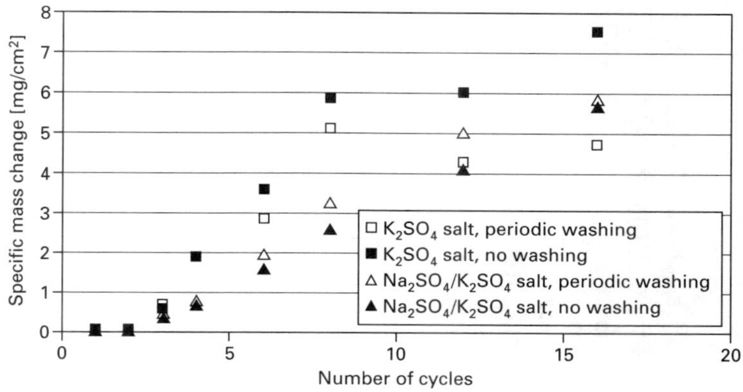

9.2 Net specific mass change data for P91 after exposure in low-velocity burner rig at 650 °C.

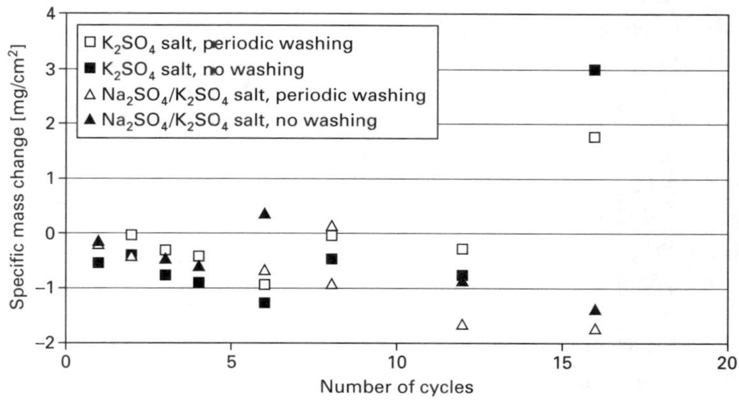

9.3 Net specific mass change data for AISI 441 after exposure in low-velocity burner rig at 850 °C.

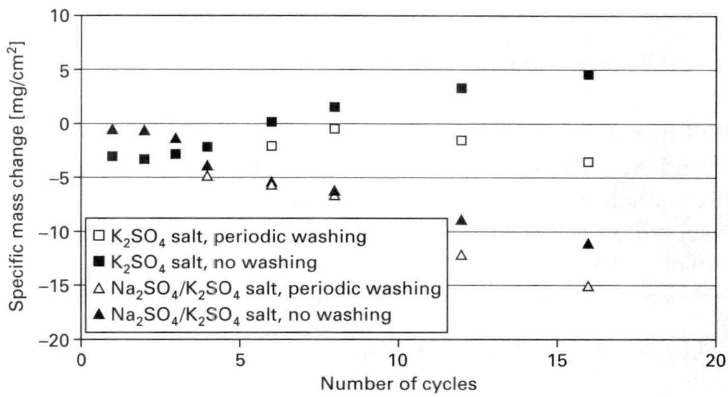

9.4 Net specific mass change data for alloy 800H after exposure in low velocity burner rig at 950 °C

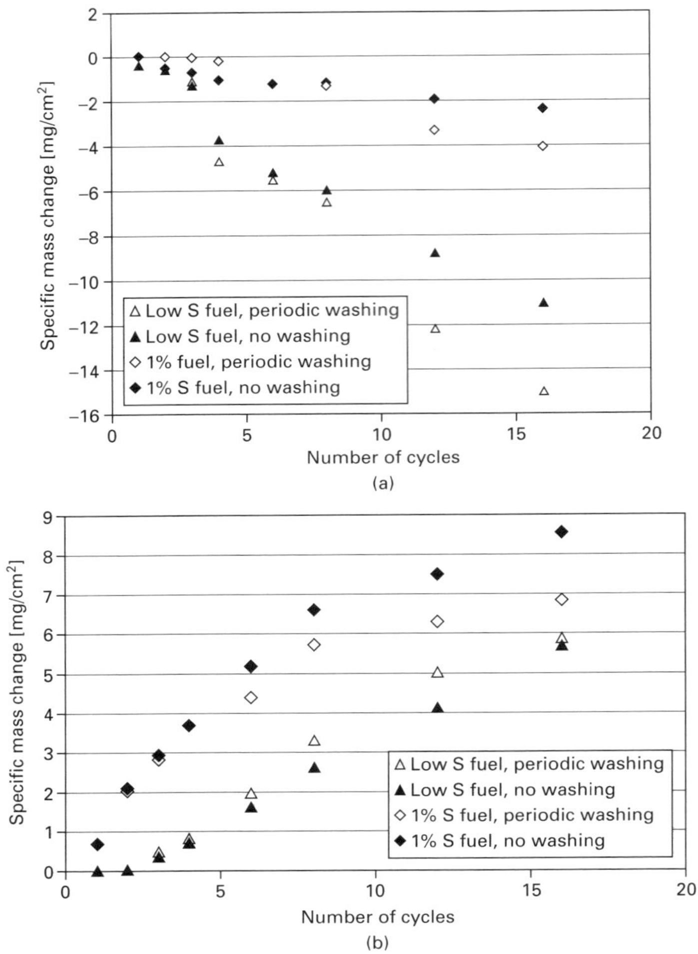

9.5 Comparison of net specific mass change data after exposure to low S and 1% S kerosene fuel in low-velocity burner rig (a) P91 at 650 °C and (b) Alloy 800H at 950 °C.

exposure in the low-velocity burner rig for 16 cycles at 850 °C. In both K_2SO_4 and $Na_2\ SO_4/K_2SO_4$ the corrosion product is damaged, consistent with the spallation inferred from the mass change data. Internal corrosion was observed on Alloy 800H after exposure in the low-velocity burner ring for 16 cycles at 950 °C for 320 h.

The recommendations arising from this investigation were as follows:

- Test pieces shall, subject to limitations in available material, be of circular section and have a domed upper surface avoiding any sharp edges.
- The test shall be defined according to the parameters – temperature

cycle, chemical composition of fuel and salt, fuel/air ratio and salt deposition rate.

- The testing duration shall be at least 300 h of accumulated hot dwell time to allow a significant oxidation of the test pieces. For more reliable results it is, however, recommended to extend the accumulated hot dwell time to 1000 h.
- Some test pieces may be washed after mass measurements. In this case the mass change prior to and after washing shall be recorded.

These recommendations were incorporated in the draft Code of Practice for cyclic oxidation testing.

9.3 High-velocity burner rig

MTU Aero Engines performed cyclic oxidation tests on MAR-M247 material in the high-velocity burner rig. The atmosphere used in this test was burnt jet fuel and the upper temperature was 1150 °C. The lower temperature in each cycle was 25 °C and the cooling of the specimens was realised by using compressed air. The total amount of six specimens tested was split into two sets, each containing three specimens. The difference in testing condition that was varied between those two sets was the dwell time at the lower temperature. It was planned to evaluate the influence of time at the lower temperature on the oxidation behaviour of the material. The detailed test conditions are given in Table 9.2.

The measured parameters were mass change and radial change for each specimen. The post-test examinations of the specimens were done by metallographic cross-section. The test results of mass change and radial change as a function of time from set 2 are given in Figs 9.6 and 9.7.

One of the two tested sets (with 18 min cooling) unfortunately cannot be taken into account for evaluating the influence of time at lower temperature on the oxidation behaviour, owing to a temperature problem in the burner

Table 9.2 Test matrix for high-velocity burner rig testing, including the total exposure time of the specimens

Specimen no.	T_{max} (°C)	T_{min} (°C)	Dwell time at T_{max} (min)	Dwell time at T_{min} (min)	Total exposure time (h)
1	1150	35	57	18	298
3					298
4					298
5	1150	35	57	3	341
6					298
7					298

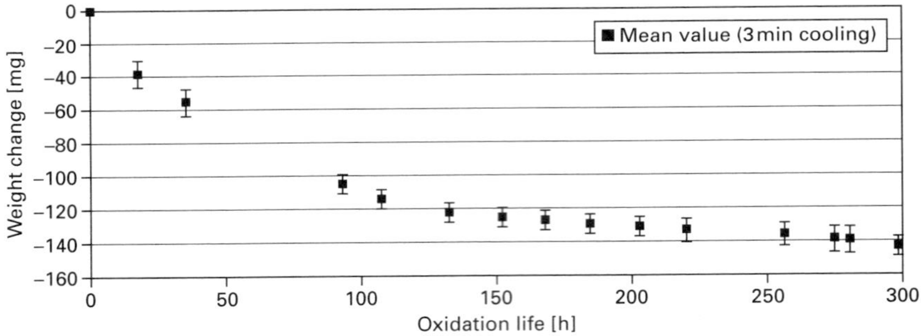

9.6 Weight change data as a function of time for the set of specimens tested with a time of 3 min cooling at lower temperature. The data points represent mean values of three specimens, the standard deviation (±σ) is given by error bars.

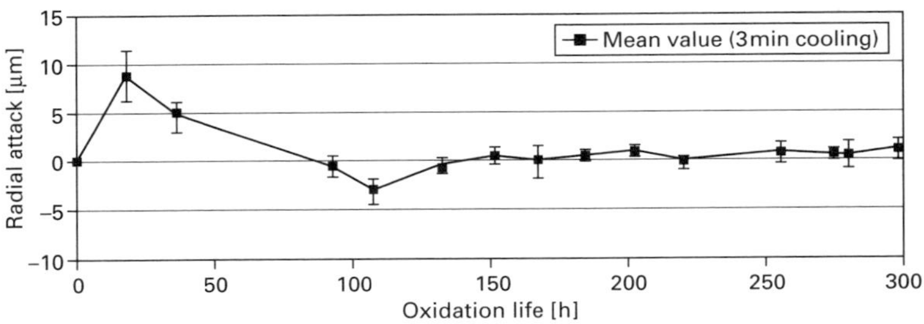

9.7 Radial change data as a function of time for the set of specimens tested with a time of 3 min cooling at lower temperature. The data points represent mean values of three specimens, the standard deviation (±σ) is given by error bars.

rig. This set of specimens was several minutes at temperatures above 1150 °C, so that a comparison with the other set is no longer possible. As a consequence, the results of the mentioned set of specimens are not shown.

Figures 9.6 and 9.7 show the test results of the set of specimens tested with 3 min cooling time. Each data point represents the mean value of three tested specimens. The standard deviation (±σ) is given by error bars. For the individual mass change data points of each specimen three replicate measurements were taken. The standard deviation for the weighing is a maximum of 0.0006 mg.

The scale growth of oxide is more difficult to measure with this kind of burner rig test compared with cyclic furnace tests. The scale growth in burner rig could not be measured by using the mass change per area method, but only relative mass change. The reason is a temperature gradient along the test bar that leads to uneven oxidation attack. Significant material loss from the beginning of the test due to evaporation of several elements of the material in the hot gas stream leads to a weight loss (Fig. 9.6) and not to a weight gain as in the cyclic oxidation test in air.

Scale growth could be measured with the radial change method. In Fig. 9.7 the increase of radius of the tested specimens within the first 50 h of testing can be seen. This is a consequence of grown Al-oxide scale. After 50h spalling occurred and the specimen radius dropped.

9.3.1 Macroscopic appearance of the MAR-M247 specimens

The macroscopic appearance of the tested specimens of the set with 3 mins at lower temperature is given in Fig. 9.8. Owing to a temperature gradient on the test bars the oxide is not uniform on the specimen. The light grey oxide is aluminium-oxide and the darker spots are mainly made of $NiAl_2O_4$. Some small NiO and CoO spots formed as well.

9.3.2 Examinations on metallographic cross-section

Table 9.3 shows the test results of the metallographic examinations of one of the specimens shown in Fig. 9.8. Twenty-four measurements were taken, uniformly distributed on the entire circumference of the specimen.

Figure 9.9 shows a typical microstructure of a MAR-M247 specimen after 300h (~300 cycles) of burner rig oxidation. This includes the heterogeneous γ'-structure, the depleted zone near the specimen surface, the grain boundary and internal oxidation and the pores. The thin oxide layer on the specimen surface is situated between the Ni-coating and the base material.

Figures 9.10 and 9.11 show some detailed scanning electron microscopy (SEM) pictures of typical oxides on the MAR-M247 surface. The specimen examined is the same as shown in Fig. 9.4. Typical oxides are Al oxide, Hf-

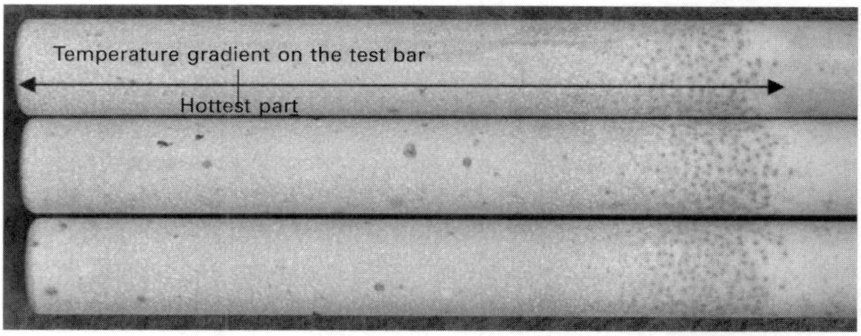

9.8 MAR-M247 specimens after about 300h of burner rig testing at 1150°C. The specimen diameter is 8mm.

Table 9.3 Results of metallographic examination of one specimen of set 2 after about 300 h of testing. Mean values of 24 measurements, plus the maximum value which is not included in the mean value

	Oxide thickness (μm)	Depth of internal oxidation (μm)	Grain boundary attack (μm)	Depth of γ'-depleted (small grains) zone (μm)	Depth of pores (μm)	Diameter of unaffected material (mm)	Radius of unaffected material (mm)
Mean values	10	17	119	121	275	7.72	3.85
Standard deviation	6	5	55	21	59	0.01	0.07
Max value (not included in mean value)	36	28	210	250	500	7.68	3.71

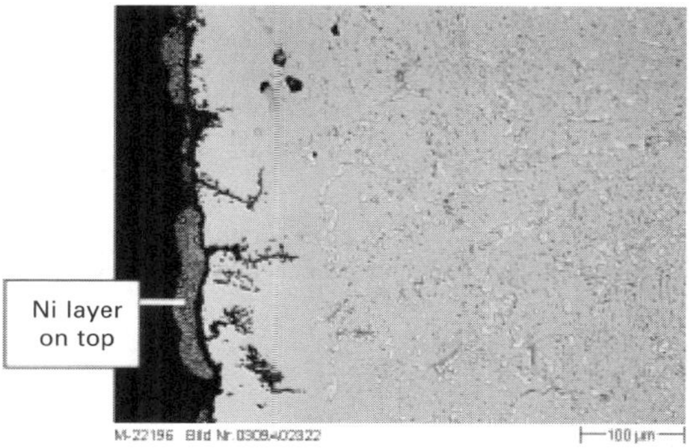

9.9 Typical microstructure of MAR-M247 after 300 h of cyclic oxidation. The outer depleted zone, the oxide layer on top of the specimen, the grain boundary and inner oxidation of the material are all visible.

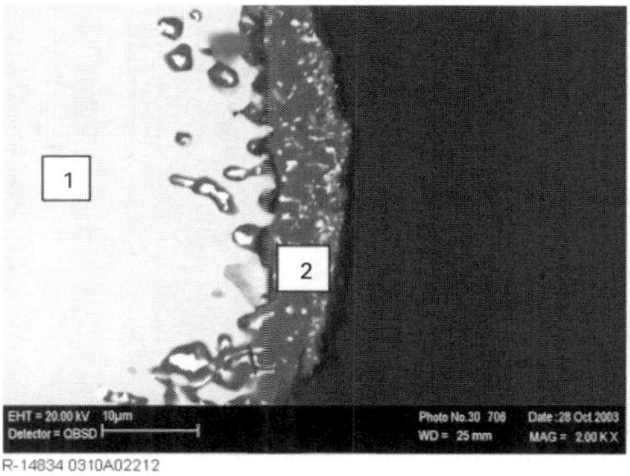

9.10 Typical oxide scale on a MAR-M247 specimen after 300 h of testing. (1) Base material. The oxide (2) is composed of Al oxide (dark) and Hf-rich oxide (light spots).

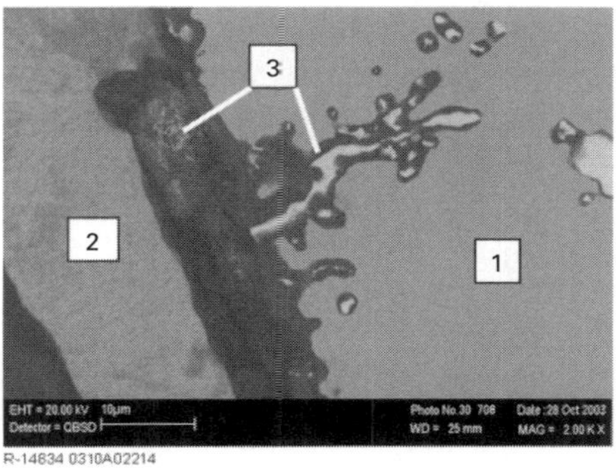

9.11 Oxide scale on MAR-M247 after 300 h of cyclic oxidation. (1) Base material, (2) Ni layer, (3) different kinds of oxides: light area: Hf-rich oxide, dark: Al oxide, oxide mixture on the left contains Al, Ni, Ti, Hf.

rich oxide, spinel ($NiAl_2O_4$) and oxides containing Al, Ni, Ti, Hf. The Al oxide is not uniform but broken into islands.

9.4 References

1. S. R. J. Saunders, J. R. Nicholls, *High Temperature Technology*, 1989, 7, 232.
2. S. R. J. Saunders, J. R. Nicholls, *High Temperature Technology*, 1995, 13, 115.

10

Thermal cycling oxidation testing in sulphidising atmospheres

C. RINALDI and L. TORRI, CESI S.p.A., Italy and
H. P. BOSSMANN, Alstom Power Ltd, Switzerland

10.1 Introduction

This chapter presents the results of thermal cycling oxidation in sulphidising atmospheres obtained within the frame of the COTEST project. The test procedures described in Chapter 4 have been applied to check for a general applicability. CESI performed experiments on the Fe-based materials P91, AISI 441 and Alloy 800H, and Alstom Power investigated the Ni-based material CM 247.

10.2 Fe-based materials

The Fe-based materials P91, AISI 441 and Alloy 800H were subjected to long dwell thermal cycling oxidation in a sulphidising-oxidising environment (0.3 vol% SO_2 and 2.5 vol% SO_2 in air). The complete test matrix is given in Table 10.1.

The furnace was equipped with a tube and crucibles in quartz for temperatures up to 900 °C. A tube and crucibles of alumina were used for higher temperatures. The temperature profiles for the tests on AISI 441 and Alloy 800H are shown in Figs 10.1 and 10.2.

Table 10.1 Test matrix

Test number	Material	Dwell time (h)		Temperature (°C)	Exposure time (h)
		Hot	Cold		
1	P91	20	4	650	300
2	P91	20	4	700	300
3	AISI 441	20	4	800	150
4	AISI 441	20	4	850	400
5	AISI 441	20	4	900	400
6	Alloy 800H	20	4	1000	180
7	Alloy 800H	20	4	1050	180

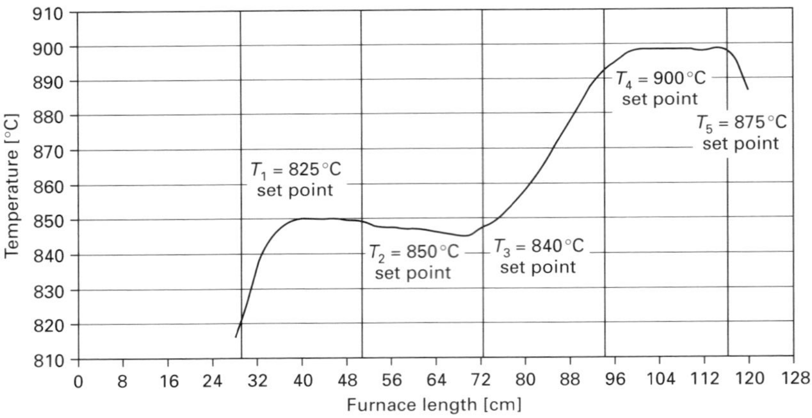

10.1 Temperature inside the furnace. Samples were tested in the regions at 850 and 900 °C.

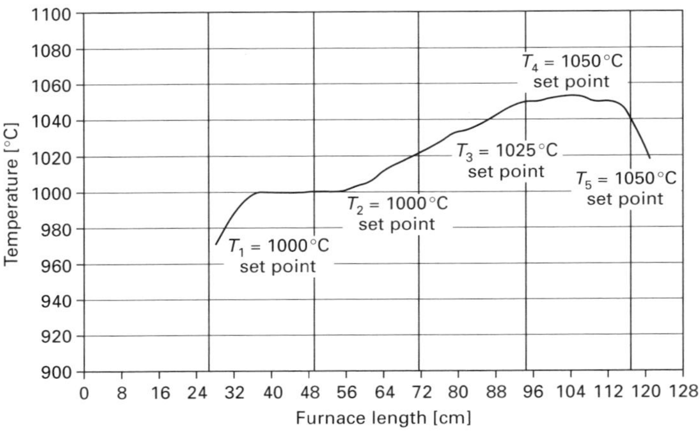

10.2 Temperature inside the furnace. Samples were tested in the regions at 1000 and 1050 °C.

In Figs 10.3 and 10.4 the results obtained on the P91 steel at 650 and 700 °C are shown in terms of mass change. Metallographic analyses on the transverse section of one specimen for each test temperature were performed. In Fig. 10.5 the results of such observations in the SEM are reported. Comparing oxide morphology with the X-ray maps, the followed can be noted:

- At 650 °C the oxide grown on the P91 is formed by a double layer: the internal layer is rich in Cr and the external layer is rich in Fe and an incipient spalling is present.
- In contrast, at 700 °C the oxide is thinner than at 650 °C and is formed only by a single layer very rich in Cr. This possibly happens because the

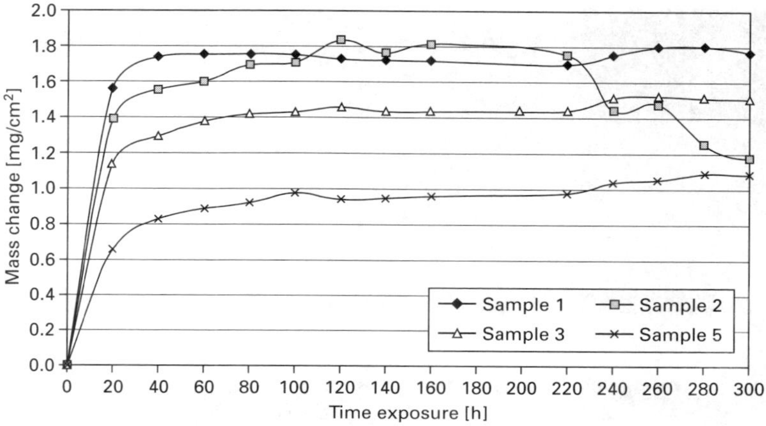

10.3 Results of the test at 650 °C on P91 steel.

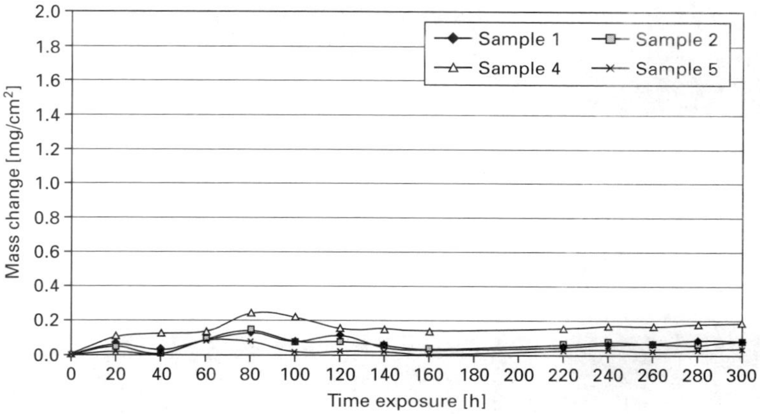

10.4 Results of the test at 700 °C on P91 steel.

diffusion kinetic of Cr at higher temperature becomes faster and enables the rapid formation of a protective and stable chromium oxide; this was observed at 700 °C also in other environments by CESI (COST programme) and by other authors [1].

In Figs 10.6–10.8 the results obtained at the three test temperatures on AISI 441 are shown in terms of mass change. AISI 441 shows that at 800 °C after only 100 hours' exposure there is a very high corrosion rate due to the phenomenon of breakaway. A comparison of Figs 10.7 and 10.8 shows that for AISI 441 the weight gain (and corrosion rate) at 900 °C is higher and faster than at 850 °C.

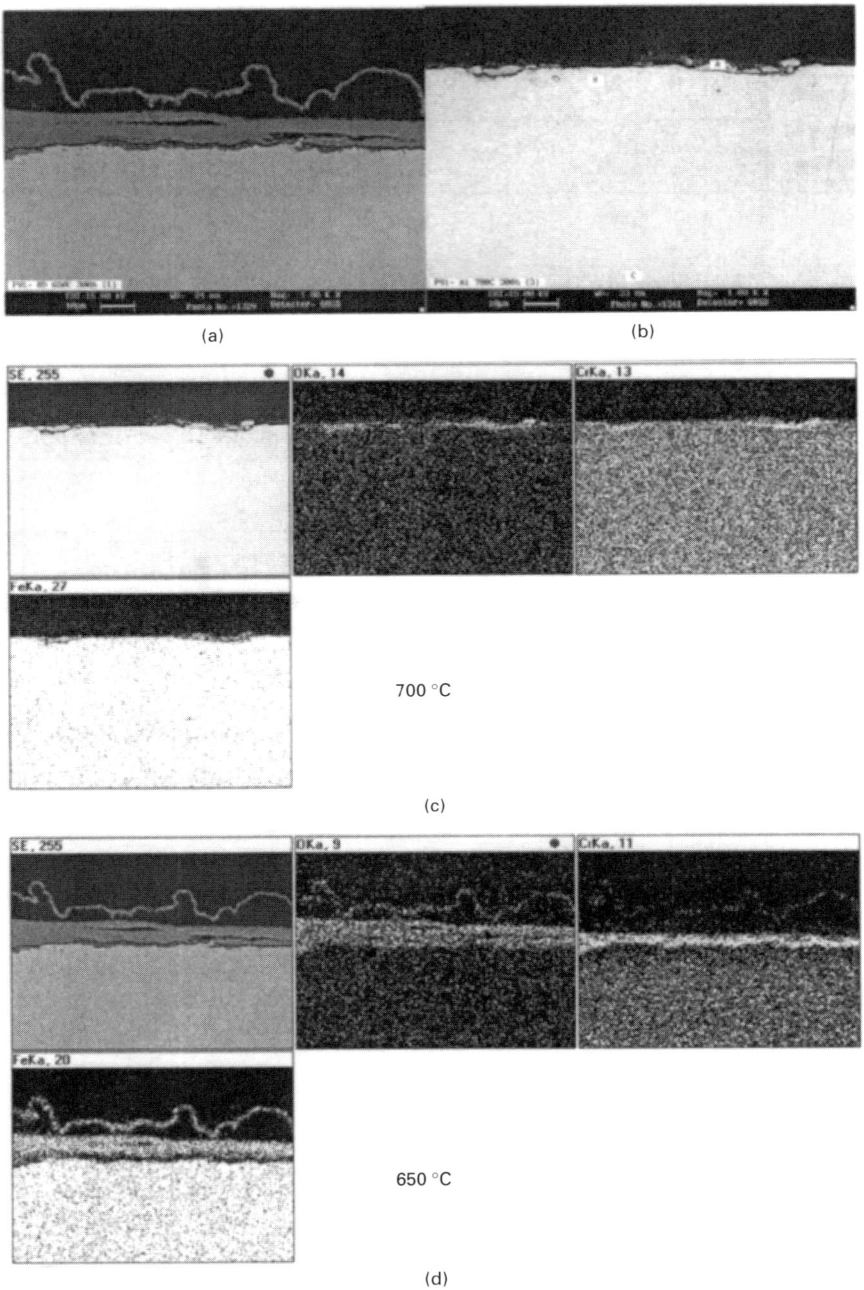

10.5 P91 SEM pictures of oxide morphology at 650 and 700 °C (top left and right) with respective energy dispersive spectrometer (EDS) maps.

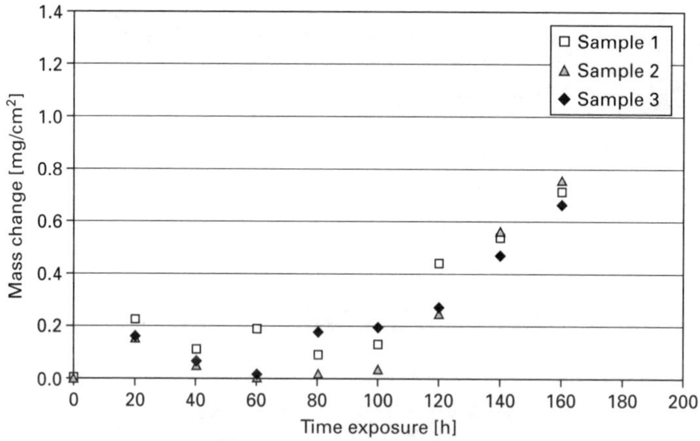

10.6 Results of the test at 800 °C on AISI 441 (breakaway).

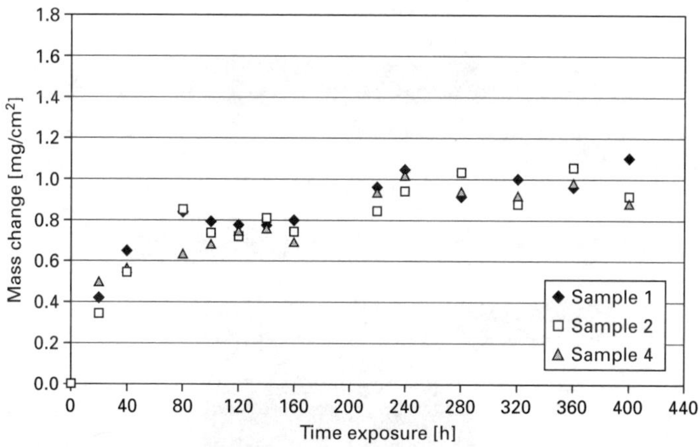

10.7 Results of the test at 850 °C on AISI 441.

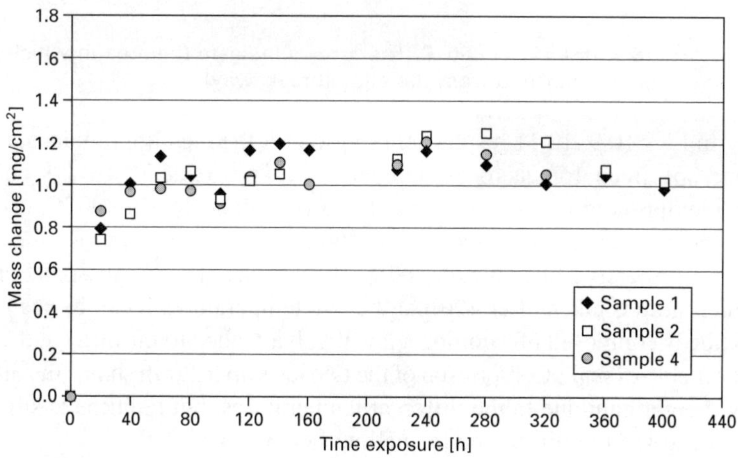

10.8 Results of the test at 900 °C on AISI 441.

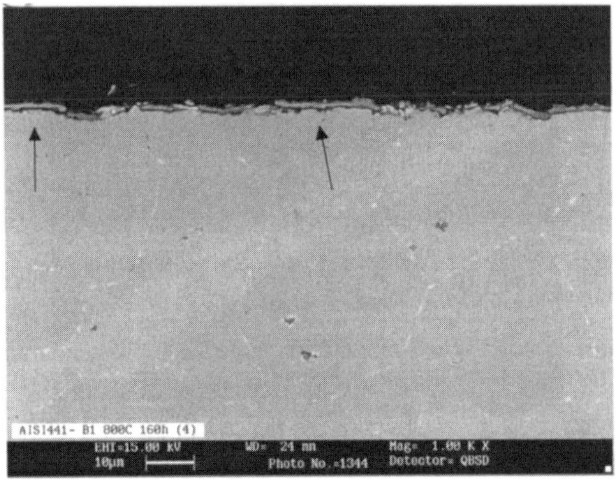

(a)

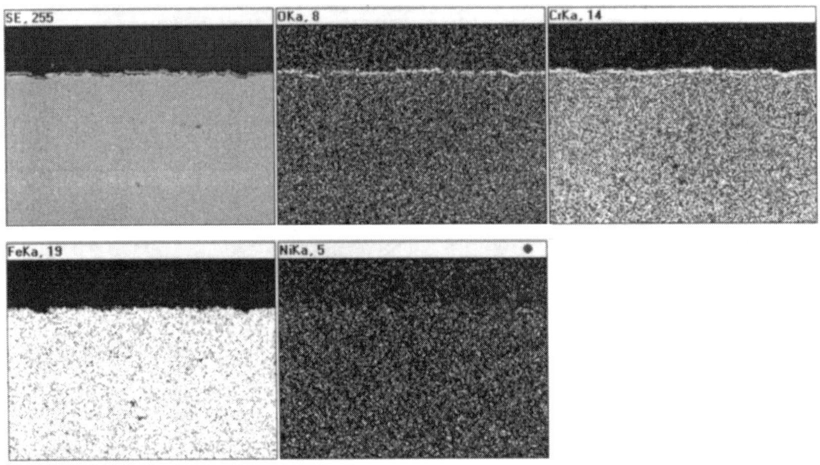

(b)

10.9 AISI 441 at 800 °C. The arrows indicate the area in which the oxide detached from the alloy (breakaway).

In Figs 10.9–10.11 results are reported for the metallographic analyses on one specimen for each test temperature. The oxide is always chromium oxide; it is compact but detaches after a very short time at 800 °C, giving rise to the breakaway phenomenon. At 850 and 900 °C, in contrast, the oxide is more protective but towards 280 h of exposure it starts to spall. From the micrographs taken after 400 h at the two temperatures it can be seen that the oxide is composed of a compact layer well attached to the alloy and a surface layer able to spall. Comparison of the two tests up to 300 h shows the difference of weight gain due to the different temperatures, but the time is slightly too short to see the difference in spalling behaviour clearly.

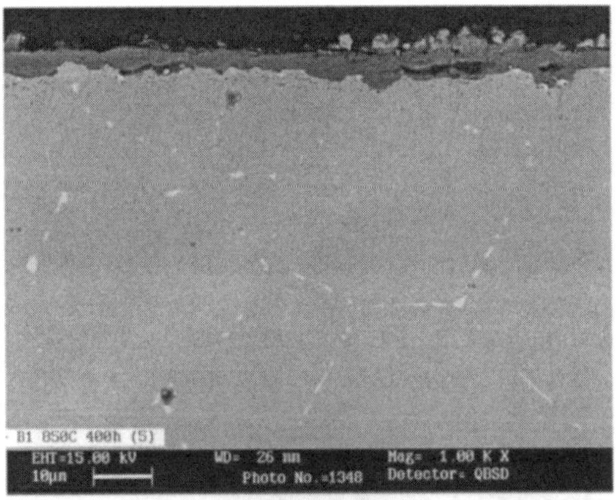

(a)

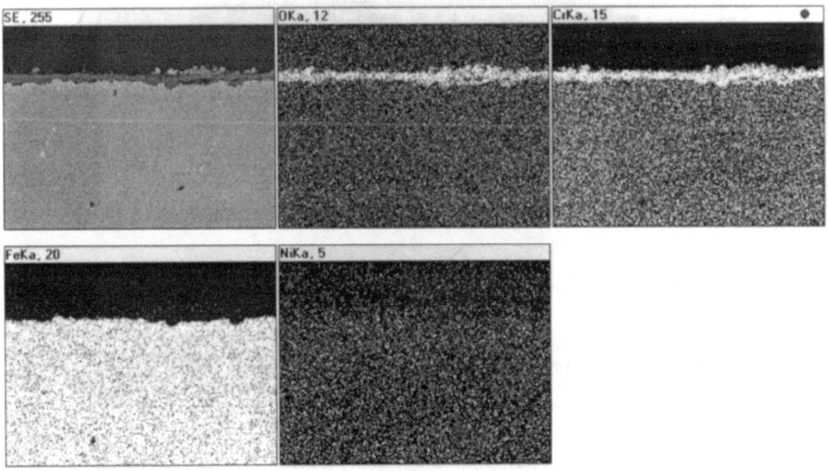

(b)

10.10 AISI 441 after 400 h exposure at 850 °C (20 cycles 20 h hot and 4 h cold).

A comparison is shown in Fig. 10.12 between the results obtained at 900 °C in a corrosive environment and in air: the two oxidation curves have the same behaviour although the presence of sulphur in the mixed gas atmosphere accelerates the weight change, as expected.

The tests on Alloy 800H were performed at 1000 and 1050 °C, with an alumina tube in the furnace; unfortunately the frequent opening of the furnace (every day) for the required cycling of samples to room temperature gave rise to tube breaking after nine cycles (test duration only half the one required in the draft code). In Figs 10.13 and 10.14 the results obtained for the three test temperatures on Alloy 800H are shown in terms of mass change. Even

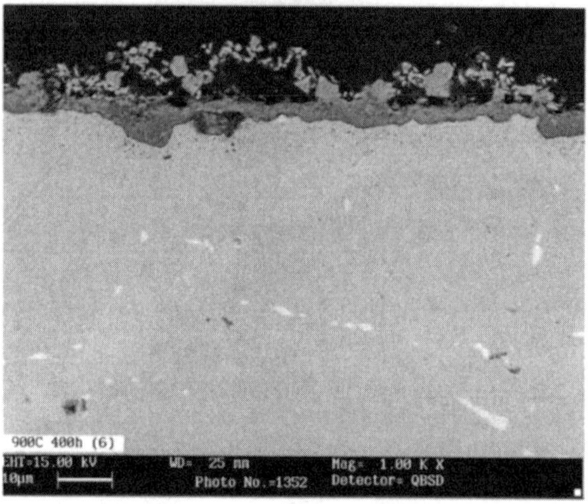

(a)

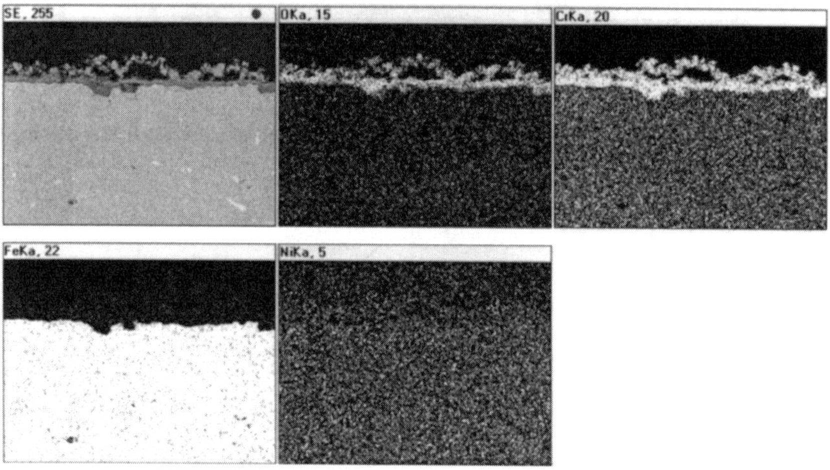

(b)

10.11 AISI 441 after 400 h at 900 °C (20 cycles of 20 h hot and 4 h cold); notice the double-layered nature of the oxide; the external surface layer is buckling and spalling.

when the test was stopped before the required duration, the difference associated with the test temperature can be seen by comparing the two plots.

In Figs 10.15 and 10.16 it is possible to see the results of metallographic analyses performed in the SEM on one specimen for each test temperature. The X-ray maps show:

- the nature of the oxide layer (one internal Cr-rich layer and a surface of a more brittle Fe and Ni-rich layer, able to spall easily);

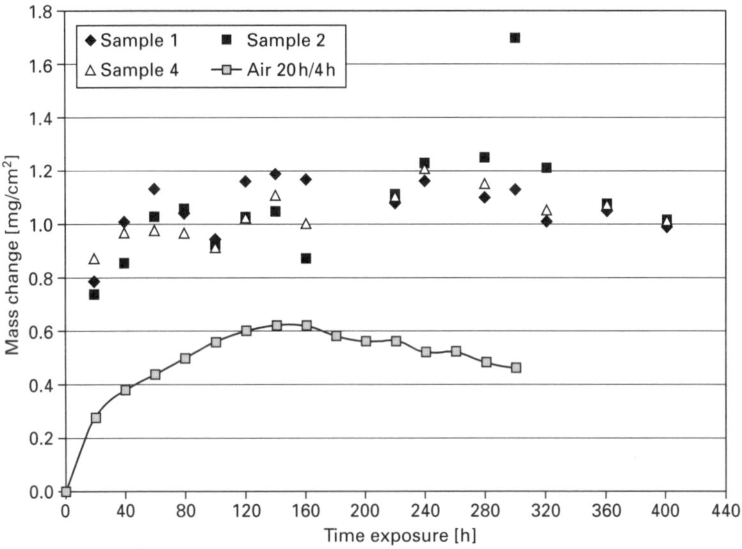

10.12 AISI 441 at 900 °C comparison between exposures in mixed gas environment and air.

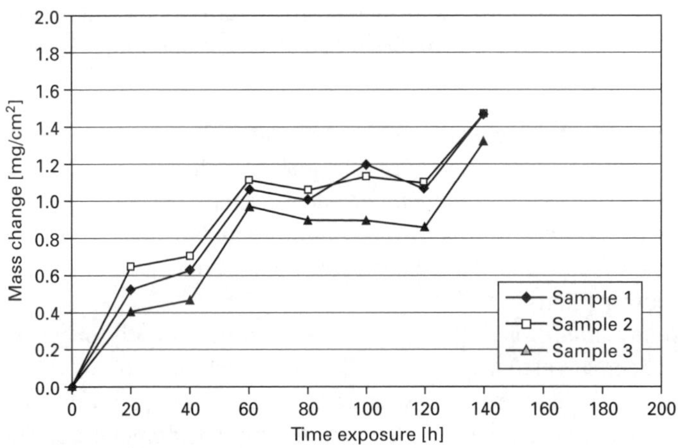

10.13 Results of the test at 1000 °C on Alloy 800.

- Si-rich precipitates rich under the oxide layer;
- internal oxidation/sulphidation.

The test results of AISI 441 and Alloy 800 have been elaborated using the procedure suggested in the draft Code of practice, in order to obtain the *n* values for the oxidation curves. In Fig. 10.17 an example is shown of the

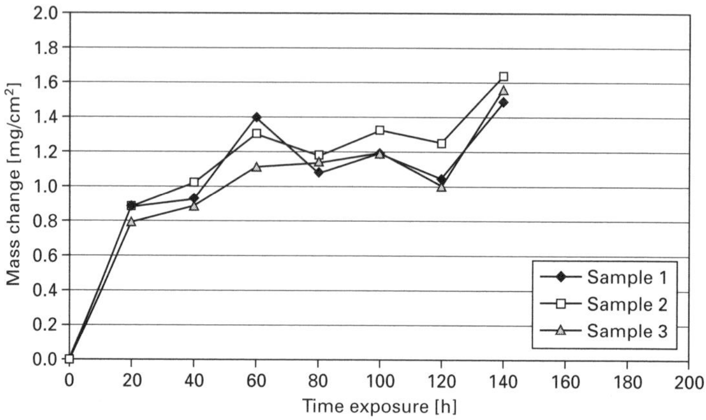

10.14 Results of the test at 1050 °C on Alloy 800.

interpolation analyses and in Table 10.2 the complete set of *b* and *n* values is reported.

10.3 Ni-based materials

The influences of (i) temperature, (ii) corrosive environment and (iii) different cycling conditions on the oxidation/corrosion behaviour of CM247 have been investigated.

- *Temperature dependence*: 900, 1000 and 1050 °C have been selected as temperatures of interest, e.g. relevant for industrial gas turbines.
- *Corrosive gas environment*: beside 'normal' lab air with approx. 1.5 wt% humidity, air with 10 wt% humidity and air with 10 wt% humidity plus 0.03 wt% SO_2 have been selected, e.g. relevant for oil combustion in power plants.
- *Cycling conditions*: the oxidation behaviour for daily or weekly cycles has been investigated. The daily cycle consisted of working days of 20 h dwell at upper temperature and 4 h dwell at room temperature; during weekends the samples stayed at upper temperature. The weekly cycle consisted of 164 h dwell at upper temperature and 4 h dwell at room temperature.

The evaluation of the oxidation/corrosion behaviour of samples tested with the weekly cycling for approximately 328 h (two cycles 164 h hot dwell time) revealed a big scatter. Therefore the exposure time of seven tests has been extended to 600 h (four cycles) and 1200 h (seven cycles). Four tests at 900 °C, which were not believed to be essential for the evaluation of the above-mentioned effects, were cancelled (Table 10.3).

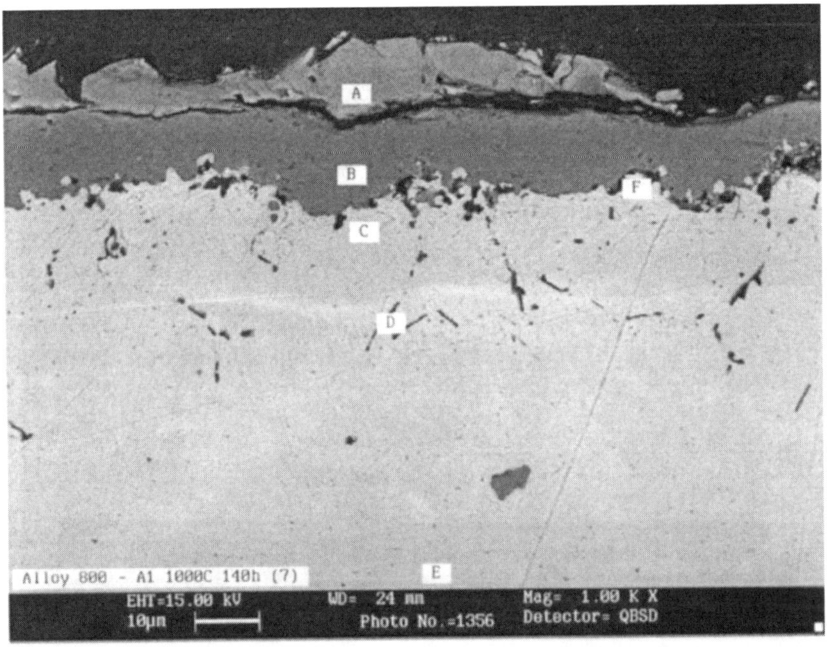

(a)

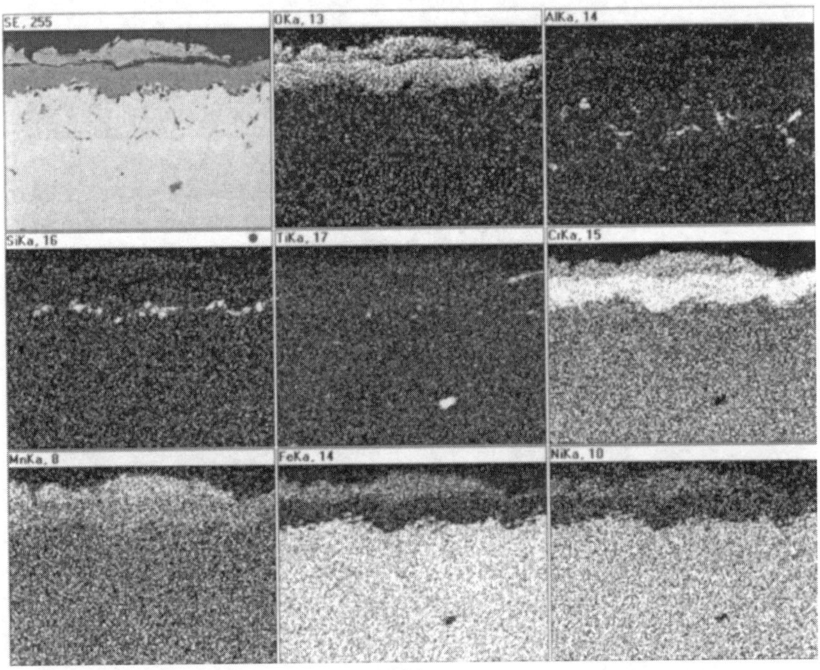

(b)

10.15 Alloy 800 at 1000 °C (180 h).

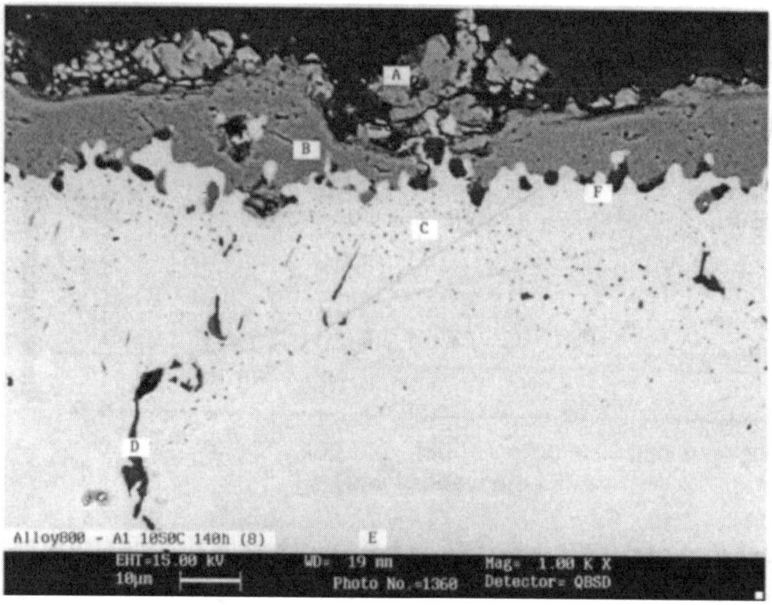

(a)

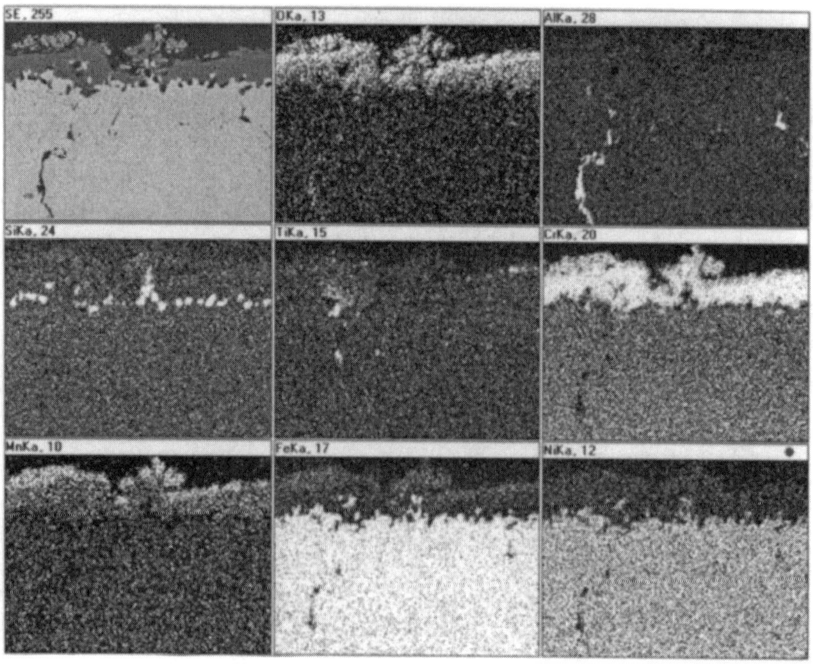

(b)

10.16 Alloy 800 at 1050 °C (180 h).

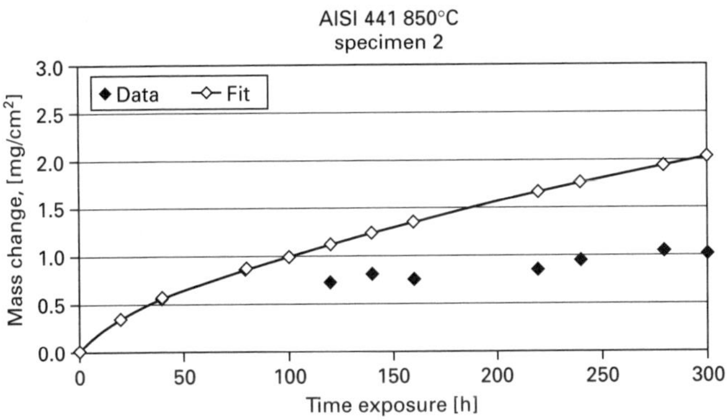

10.17 Example of data analysis applied to one test on AISI 441 at 850 °C (data from Fig. 10.7).

Table 10.2 Interpolation results

Material	Temperature (°C)	Specimen	*b*	*n*	Time (h)
AISI 441	850	1	0.4944	2.022	100
AISI 441	850	2	0.6528	1.532	100
AISI 441	850	3	0.1720	5.814	100
AISI 441	900	1	0.3279	3.049	60
AISI 441	900	2	0.2730	3.663	80
Alloy 800H	1000	1	0.5259	1.901	100
Alloy 800H	1000	2	0.3748	2.668	100
Alloy 800H	1000	4	0.5483	1.824	120
Alloy 800H	1050	1	0.1837	5.443	80
Alloy 800H	1050	2	0.2415	4.141	60
Alloy 800H	1050	4	0.2683	3.727	100

The cycling boundary conditions have been adjusted to the heating and cooling behaviour of the samples. A dummy sample (same material, size and sample holder) with a spot-welded thermocouple on top of the surface was used to determine the hot and cold dwell time according to the developed guidelines. The defined start of the hot dwell time (97% of the desired temperature) has been reached within 10–13 min, depending on the different maximum temperature (Fig. 10.18). The cold dwell time is defined as $T < 50\,°C$. This has been achieved within 25.0–25.3 min. Based on this set-up the given hot dwell time has to be reduced by 0.2 h to 159.8 h for weekly cycling and 19.8 h for daily cycling; the corresponding cold dwell time has to be reduced by 0.4 h to 3.6 h for weekly and daily cycling.

In Fig. 10.19 the net mass change for weekly cycling is shown for the three different environments, and in Fig. 10.20 the corresponding data are shown for daily cycling. The big scatter for duplicate samples was not expected

Table 10.3 Overview of the test conditions used for exposure of MAR-M247

T(°C)	Atmosphere (hot)	Dwell time (h) Hot	Cold	Exposure time (h)	Status
900	lab humidity	20	4	300	done
1000	lab humidity	20	4	300	done
1050	lab humidity	20	4	300	done
900	lab humidity	164	4	1200	done
1000	lab humidity	164	4	1200	done
1050	lab humidity	164	4	1200	done
900	10% humidity	20	4		cancelled
1000	10% humidity	20	4	300	done
1050	10% humidity	20	4	300	done
900	10% humidity	164	4		cancelled
1000	10% humidity	164	4	600	done
1050	10% humidity	164	4	600	done
900	10% humidity +300 ppmSO_2	20	4		cancelled
1000	10% humidity +300 ppmSO_2	20	4	300	done
1050	10% humidity +300 ppmSO_2	20	4	300	done
900	10% humidity +300 ppmSO_2	164	4		cancelled
1000	10% humidity +300 ppmSO_2	164	4	600	done
1050	10% humidity +300 ppmSO_2	164	4	600	done

and was, therefore, investigated in detail. The top surface view of the oxidised samples showed locally different attack (Fig. 10.21). The cross-section revealed preferred attack at inter-dendritic regions (Fig. 10.22), which is a typical microstructure of MAR-M24. The comparison of corrosive environments by mass change curves showed for air plus 10% humidity no difference between 1000 and 1050 °C, but for lab air a significant difference in oxide growth (Fig. 10.19a, b) and for air plus H_2O plus SO_2 a significant difference in oxide spallation (Fig. 10.19b, c). This behaviour cannot be confirmed by microstructural evaluation, where the temperature also seems to have a significant effect for air plus 10% H_2O (Fig. 10.23).

For oxidation in lab air, the daily and weekly cycling can be compared (Fig. 10.24). While at 900 °C almost no difference can be seen (protective oxide growth without spallation), there are significant mechanism changes at 1000 and 1050 °C (oxide scale spallation with subsequent internal oxidation).

To evaluate the characteristic figures for oxidation, the net mass changes are shown in a log–log plot (Fig. 10.25). Some of the regression lines are less confident, especially those of high temperature and daily cycling. For weekly

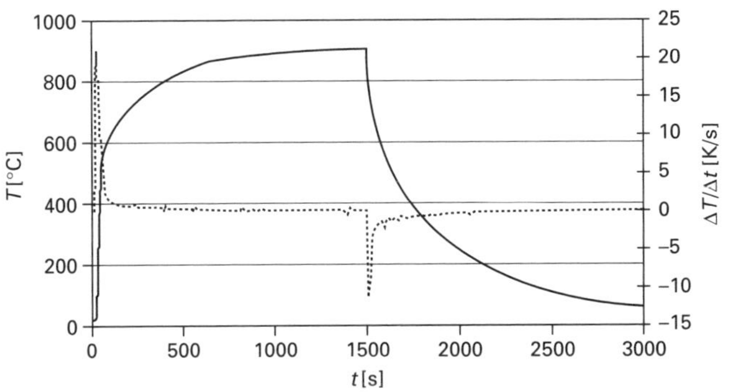

(a) Heating up and cooling down of standard oxidation samples at 900 °C

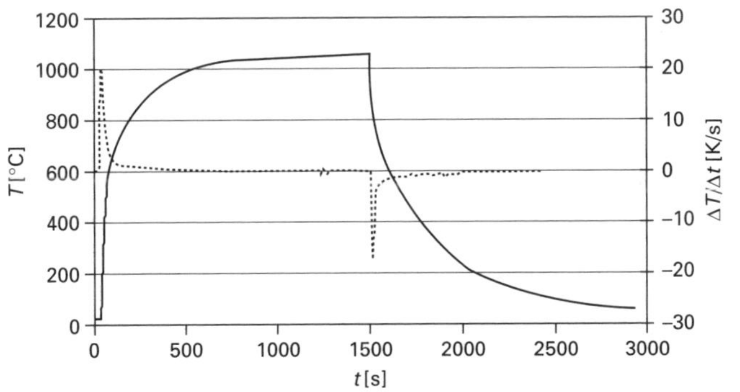

(b) Heating up and cooling down of standard oxidation samples at 1050 °C

10.18 Boundary conditions of cycling for MAR-M247 samples in ceramic holder. Left *y*-axis and solid curve: temperature. Right *y*-axis and dashed curve: heating and cooling rate.

cycling, no noticeable spallation has been observed by mass change curves. Independent of confidence of the correlation, all data of the analysis are summarised in Table 10.4. The growth rate *K* and the exponent of the oxidation power law are shown in an Arrhenius plot, revealing that the scatter of both parameters decreases with increasing temperature (Fig. 10.26).

The main results are as follows:

- The lowest oxidation without relevant spallation has been found for weekly cycling in lab air, considering the temperature range from 900 to 1050 °C.
- Daily cycling is more detrimental than weekly cycling in lab air as well as in a corrosive environment: oxide scale spallation and internal oxidation increase with increasing temperature.
- The oxidation behaviour is dependent on the type of environment: oxide scale spallation and internal oxidation increase with increasing H_2O and even more with H_2O/SO_2 in the environment.

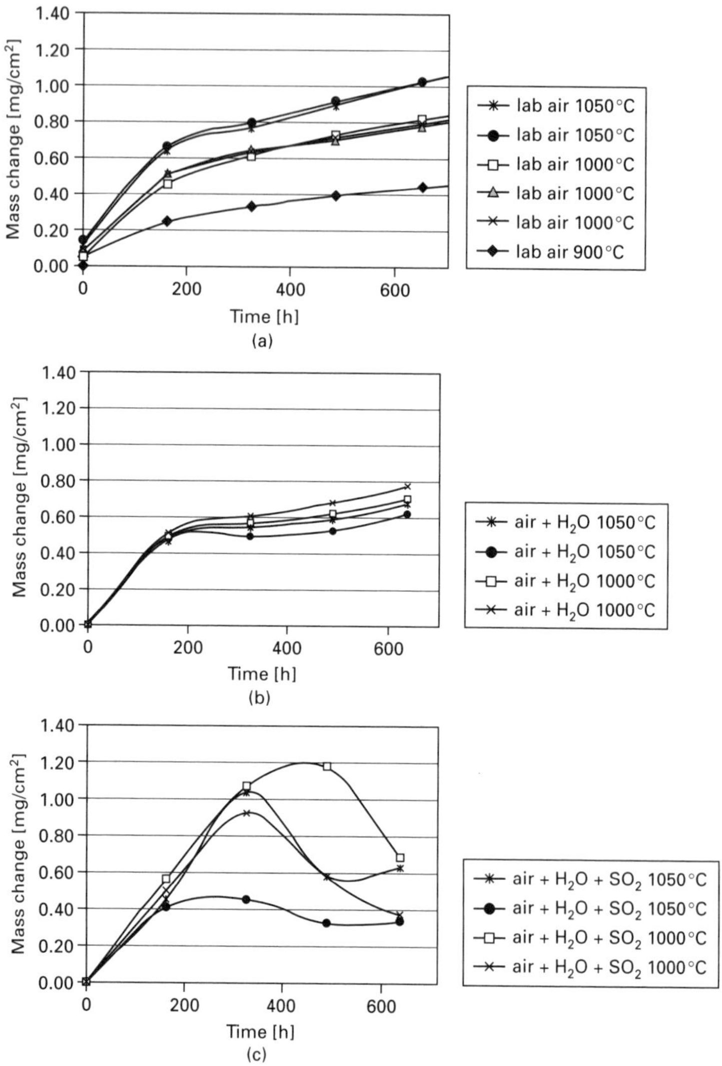

10.19 Net mass changes of MAR-M247 for *weekly* cycles: (a) lab air; (b) air with 10 wt% H_2O; (c) air with 10 wt% H_2O and 0.03 wt% SO_2 (164 h hot dwell time and 4 h cold dwell time).

- At higher temperatures MAR-M247 becomes more sensitive for environmental and cycling parameters.
- The inhomogeneous microstructure of MAR-M247 has resulted in a big scatter of data. The evaluation according to the Code of Practice requires some adaptation for this material and the boundary conditions of cycling and environment.

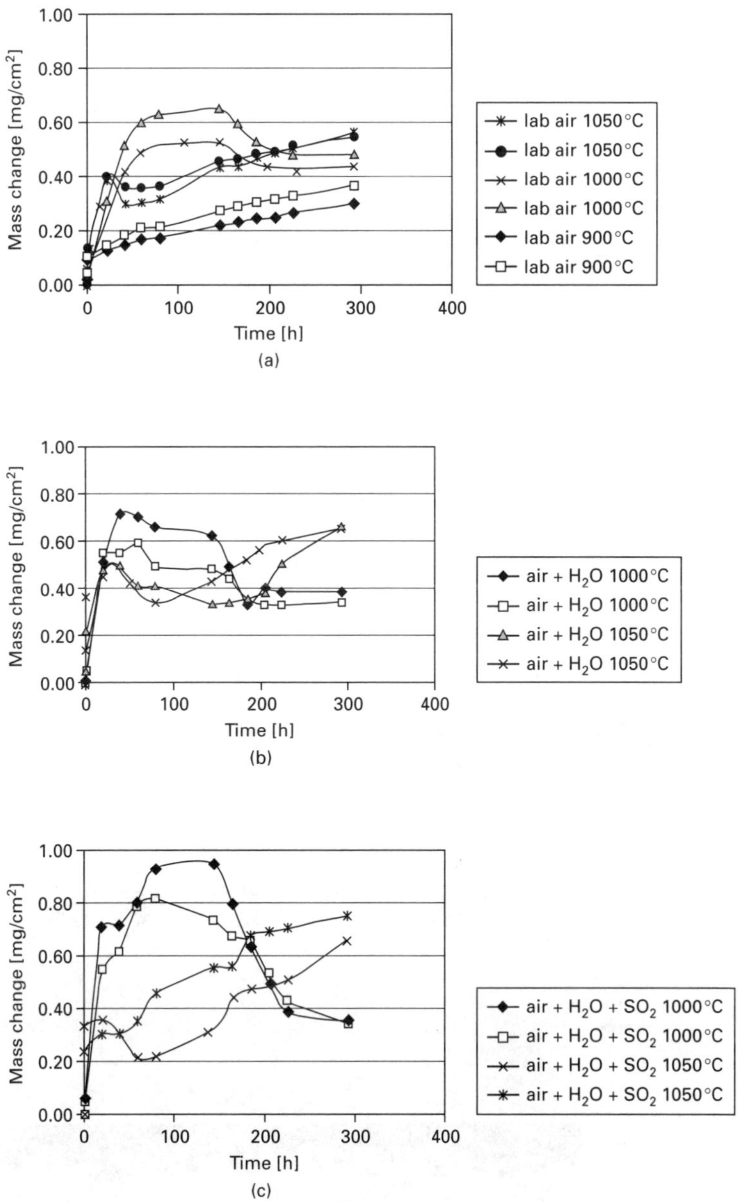

10.20 Net mass changes of MAR-M247 for *daily* cycles: (a) lab air; (b) air with 10 wt% H_2O; (c) air with 10 wt% H_2O and 0.03 wt% SO_2 (20 h hot dwell time and 4 h cold dwell time).

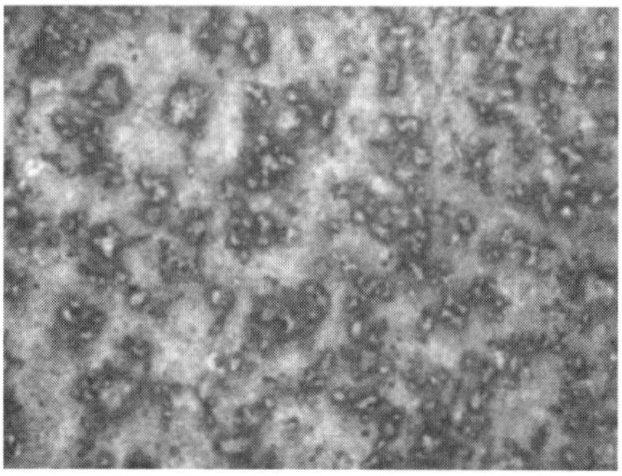

10.21 Top surface view of specimens exposed in lab air, daily cycle. Oxidation reflects microstructure of the surface with preferred attack at inter-dendritic regions. Surface area = 2.1 mm × 1.6 mm.

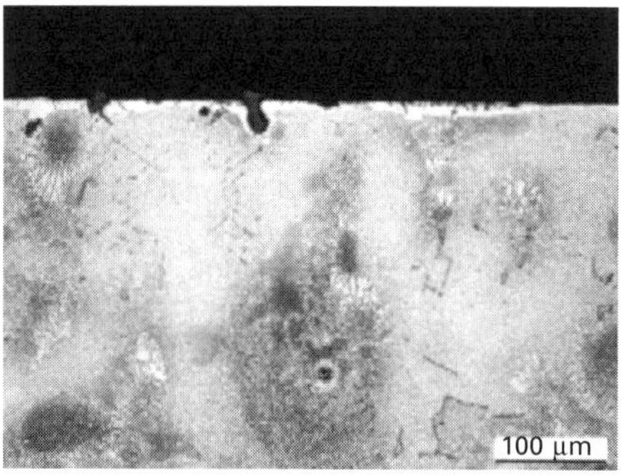

10.22 Cross-section of MAR-M247, weekly cycling at 1000 °C in air with 10% humidity. Oxidation reflects microstructure of the surface with preferred attack at inter-dendritic regions.

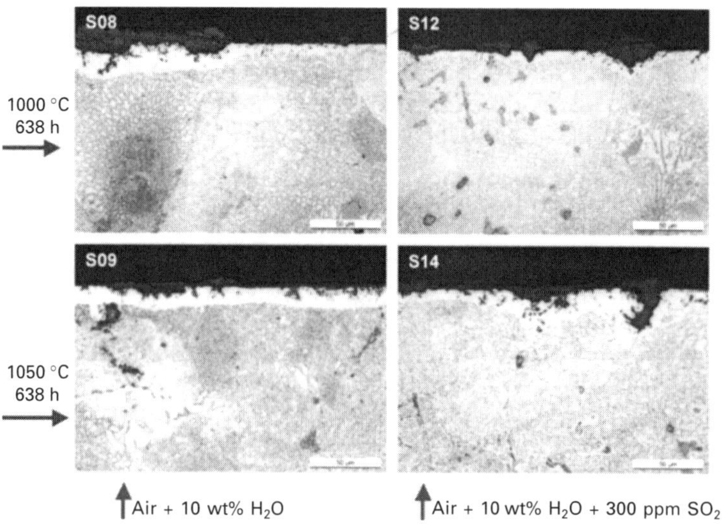

10.23 Microstructural evaluation of temperature and environmental effect.

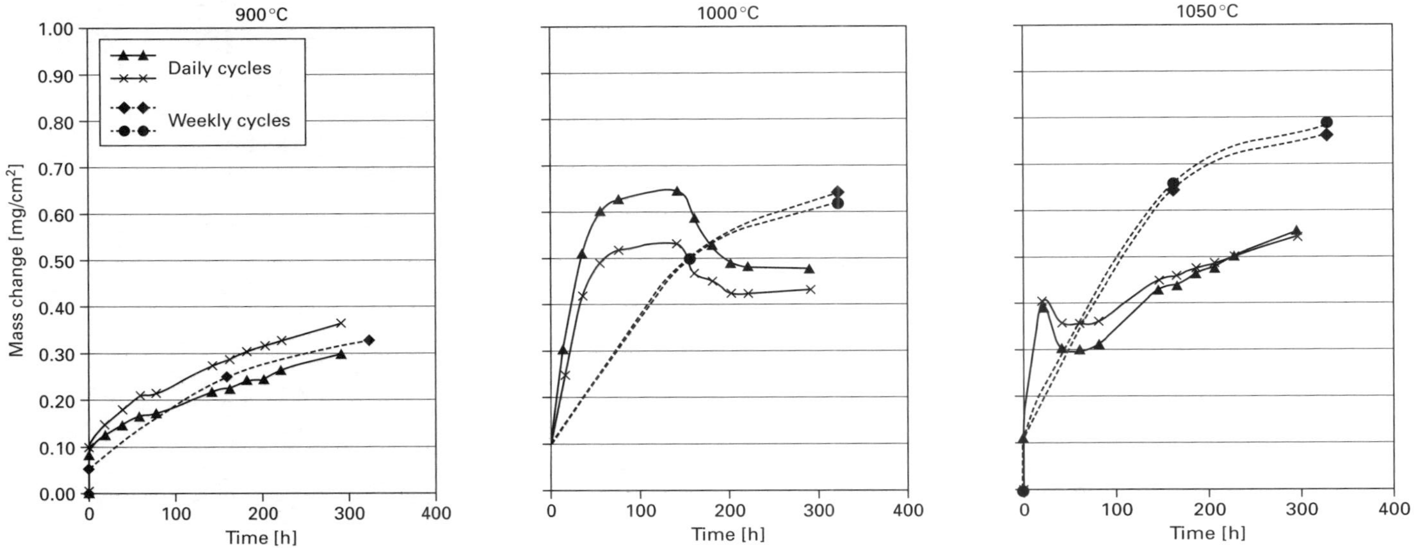

10.24 Comparison of net mass changes of daily and weekly cycling for MARM247 oxidised in lab air.

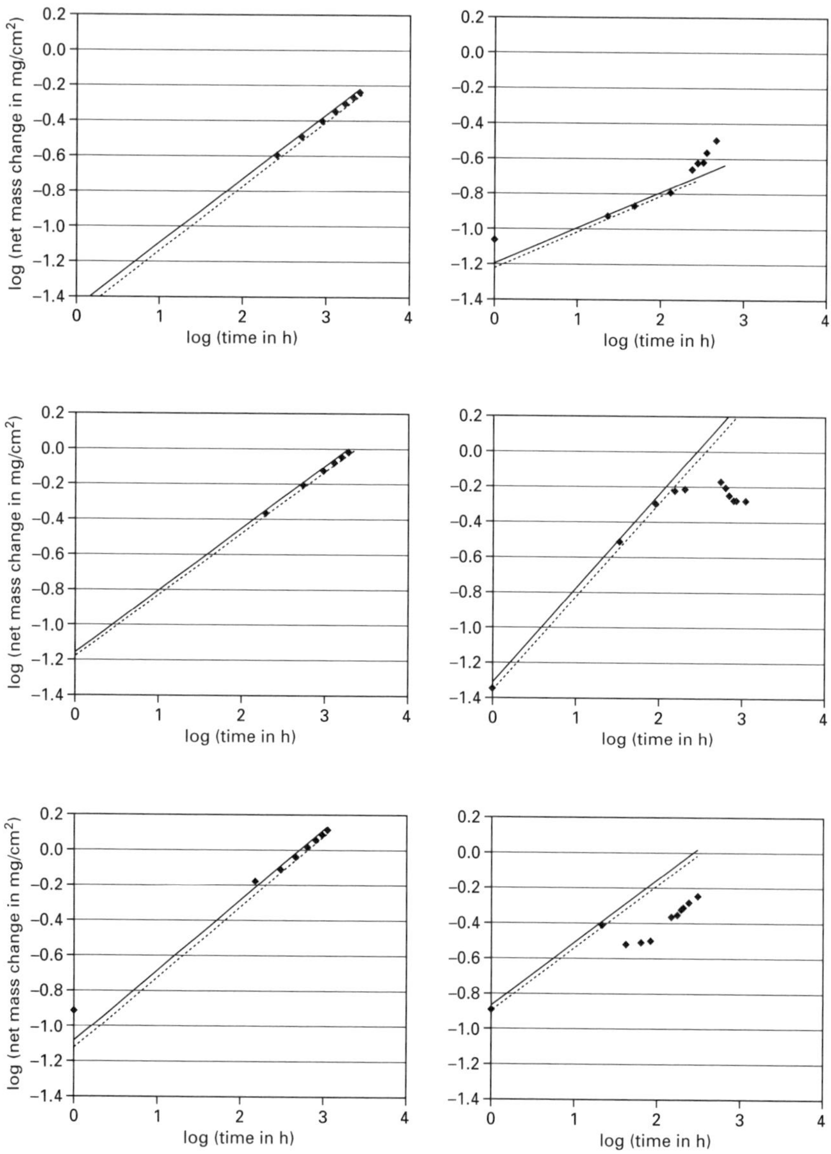

10.25 Log–log plot for the comparison of weekly cycling (left column) and daily cycling (right column) for MAR-M247 oxidised in lab air (a) 900 °C; (b) 1000 °C; (c) 1050 °C.

Table 10.4 Evaluation of MAR-M247 oxidised in lab air daily and weekly cycling

	Temperature (°C)					
	900 °C		1000 °C		1050 °C	
Daily						
K	7.4×10^{-5}	7.2×10^{-6}	4.5×10^{-3}	7.8×10^{-3}	3.7×10^{-3}	3.4×10^{-3}
n	3.4	4.3	1.8	1.6	2.8	2.8
t onset (h)	81	81	61	61	21	21
N onset (cycles)	5	5	4	4	2	2
m onset (mg/cm^2)	0.21	0.17	0.49	0.60	0.40	0.39
m (300 h) (mg/cm^2)	0.33	0.24	1.24	1.82	1.08	1.04
m (t onset, 300 h) (mg/cm^2)	–0.12	–0.07	–0.74	–1.22	–0.67	–0.65
m (1000 h) (mg/cm^2)	0.48	0.32	2.39	3.91	1.65	1.60
m (t onset, 1000 h) (mg/cm^2)	–0.26	–0.15	–1.90	–3.31	–1.25	–1.21
Weekly						
K	2.2×10^{-4}	8.6×10^{-4}	7.5×10^{-4}	8.3×10^{-4}	1.6×10^{-3}	1.6×10^{-3}
n	2.4	2.6	3.2	2.8	2.5	2.9
t onset (h)						
N onset (cycles)						
m onset (mg/cm^2)						
m (300 h) (mg/cm^2)	0.33	0.62	0.64	0.63	0.78	0.80
m (t onset, 300 h) (mg/cm^2)						
m (1000 h) (mg/cm^2)	0.55	0.98	0..94	0.97	1.26	1.21
m (t onset, 1000 h) (mg/cm^2)						

Note: The definition of parameter and the method of evaluation is given in the Code of Practice; K and n determined based on net mass change in mg/cm^2 and time in h.

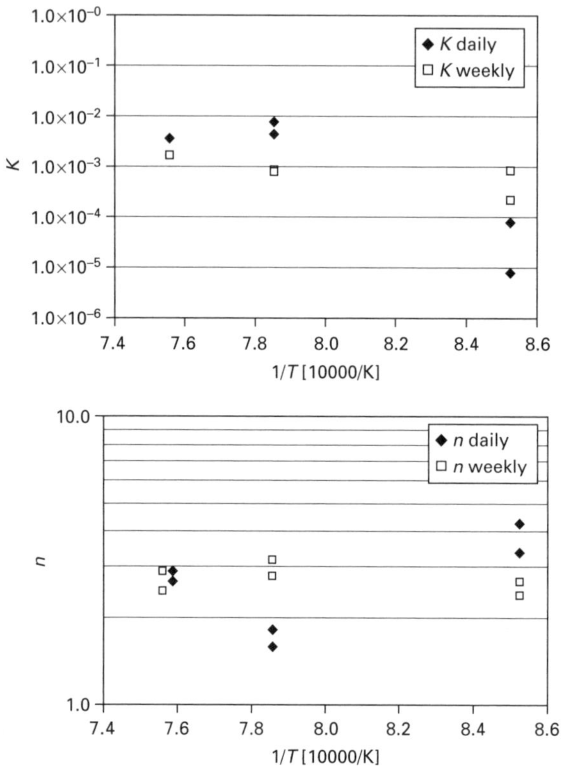

10.26 Arrhenius plot of growth rate (*K*) and exponent (*n*) of the oxidation power law for daily and weekly cycling in lab air. *K* and *n* are determined based on net mass change in mg/cm^2 and time in h.

10.4 Reference

1. N. Nishimura, N. Komai, Y. Hirayama and F. Masuyama 'Japanese experiences with steam oxidation on advanced heat resistant steel tubes in power boilers', Workshop on Scale Growth and Exfoliation in Steam Plant at NPL Teddington, London, September 2003.

11
Thermal cycling oxidation testing under deposits

M. MÄKIPÄÄ, VTT, Finland

11.1 Introduction

The contribution of VTT to COTEST mainly concerns the objective of obtaining information to ascertain whether the Code of Practice is suitable for materials selection in which additional corrosive species are acting, i.e. sulphidising-oxidising, waste incineration atmospheres and deposits. VTT's specific experience here concerns materials selection for heat exchanger surfaces that are exposed to ash deposition and temperature cycles, e.g. due to soot-blowing cycles in the superheater area of boilers using biomass or refuse as fuel.

The objective above implies that test methodologies should be defined that are similar to that in purely oxidising environments but still simple enough to be accepted in general use. Which parameters are possible/reasonable in cyclic oxidation testing were checked by the partners, resulting in a limited set of test parameters, e.g. materials to be tested at each test temperature, sample shapes and handling, upper and lower dwell times, heating and cooling rates. Additional corrosive species introduce a number of new test parameters of a kinetic type that are difficult to handle even in static corrosion testing, as experienced previously in developing discontinuous corrosion testing.

11.2 Definition of suitable test conditions

In order to obtain a meaningful but still manageable experimental test matrix in complex corrosion testing, a reduced (compared with testing in air) number of cyclic-oxidation test parameters was applied. The number of test parameters was reduced by adopting medium values of those parameters which were applied in air oxidation testing.

In the case of deposit testing, further elaboration of test conditions was considered to be necessary. Modelling calculations on the heating behaviour of the proposed alkali sulphate-alkali chloride mixture were performed by VTT using the FACT Sage database (www.factsage.com). A complex sequence of solid solution formation, partial melting and compositional changes in melt and solid solutions was predicted to occur during heating. During the high-temperature exposure dwell time, mixed gas atmosphere–deposit reactions will change the deposit mixture phase composition owing to the non-equilibrium conditions prevailing in the test conditions. Subtle changes in deposit application method, gas flow rates and flow geometry, sample material and supporting method may cause appreciable variation in corrosion conditions.

To exclude partial melting of the deposit, a test parameter that is probably most difficult to control in temperature cycling conditions, a single salt composition (K_2SO_4) having a melting point higher than the highest test temperature (1000 °C) was chosen to be used in all deposit corrosion tests in Work Package (WP5B). Test conditions to be varied were chosen as the deposit application method, upper dwell time and test atmosphere. The lower dwell time remained fixed.

11.3 Deposit testing (WP5B)

The test matrix is shown in Table 11.1. At 650 °C, upper dwell time 20 h, the number of cycles was 15 resulting in a total dwell time of 300 h. At 850 °C, upper dwell time 8 h, the number of cycles was 28, resulting in a total dwell time of 224 h.

11.3.1 Test results

Complete deposit test results are shown in Table 11.2. Three replicate weight measurements for each sample (crucible/rod/sample, crucible without sample and sample after light brush) were made. The gross mass change was defined as mass change of sample after light brushing and spall. In the actual test situation 'spall' means deposit film, not real oxide spall. Mass gain of samples embedded in deposits (deposit method burial) results from the crystals stuck on the sample surfaces/edges, i.e. material that did not fall during light brushing.

Table 11.1 Test matrix

Test no.	Material	Upper dwell	Environment	Deposit method	Deposit renewal interval
1	P91	20 h/650 °C	air + deposit	burial	after every 4 cycles
2	P91	20 h/650 °C	air + deposit	burial	after every 8 cycles
3	P91	20 h/650 °C	air + deposit	spray	after every 4 cycles
4	P91	20 h/650 °C	air + deposit	spray	after every 8 cycles
5	441	8 h/850 °C	air + deposit	spray	after 14 cycles
6	441	8 h/850 °C	air + deposit	dipping	after 14 cycles
7	441	8 h/850 °C	wet air + deposit	spray	after 14 cycles

Table 11.2 Test results

	Material	Deposit	Gross mass change (mg/cm²)	Net mass change (mg/cm²)	Spall (mg)
Test 1	P91	burial	0.13	0.13	
	P91	burial	0.25	0.25	
	P91	burial	0.26	0.26	
Test 2	P91	burial	0.00	0.00	
	P91	burial	0.06	0.06	
	P91	burial	0.21	0.21	
Test 3	P91	spray	1.72	1.17	2.45
	P91	spray	2.05	1.28	3.50
	P91	spray	2.40	1.45	4.25
Test 4	P91	spray	3.31	2.78	2.48
	P91	spray	4.21	3.36	3.76
	P91	spray	3.43	2.65	3.52
	P91	ref. *	0.69	0.69	0.00
Test 5	441	spray	0.96	0.90	0.27
	441	spray	0.71	0.68	0.14
	441	spray	1.57	1.39	0.78
Test 6	441	dip	0.35	0.16	0.89
	441	dip	0.62	0.22	1.81
	441	dip	0.76	0.40	1.64
	441	ref. *	0.36	0.36	0.00
Test 7	441	spray	–0.02	–0.06	0.19
	441	spray	–0.15	–0.16	0.05
	441	spray	–0.15	–0.17	0.08
	441	ref. *	0.36	0.35	0.05

* Reference sample without deposit

Deposit application by burial resulted, in general, in low oxide scale growth rates (Tests 1 and 2). However, net mass change measured for replicate specimens of P91 was highly variable owing to localised deposit corrosion attack; see Fig. 11.1a. Deposit application by dipping and spraying resulted in moderate variation between the net mass changes of replicate specimens of both alloys (Tests 3 to 6). In the case of P91, some localised corrosion attack was observed (Fig. 11.1b). Localised corrosion may explain the higher mass change observed in Test 4 (deposit renewal after every four cycles) as compared with that in Test 3 (deposit renewal after every eight cycles), Figs 11.1c and d.

Deposit application using the dipping method was difficult but did seemingly also result to a moderate variation in the measured mass change as observed in Test 6. However, the oxide scale formed on the material AISI 441 was in each case rather thin (see Fig. 11.2, implying that measured mass change variation may be in part attributed to the dipping method. Spraying is

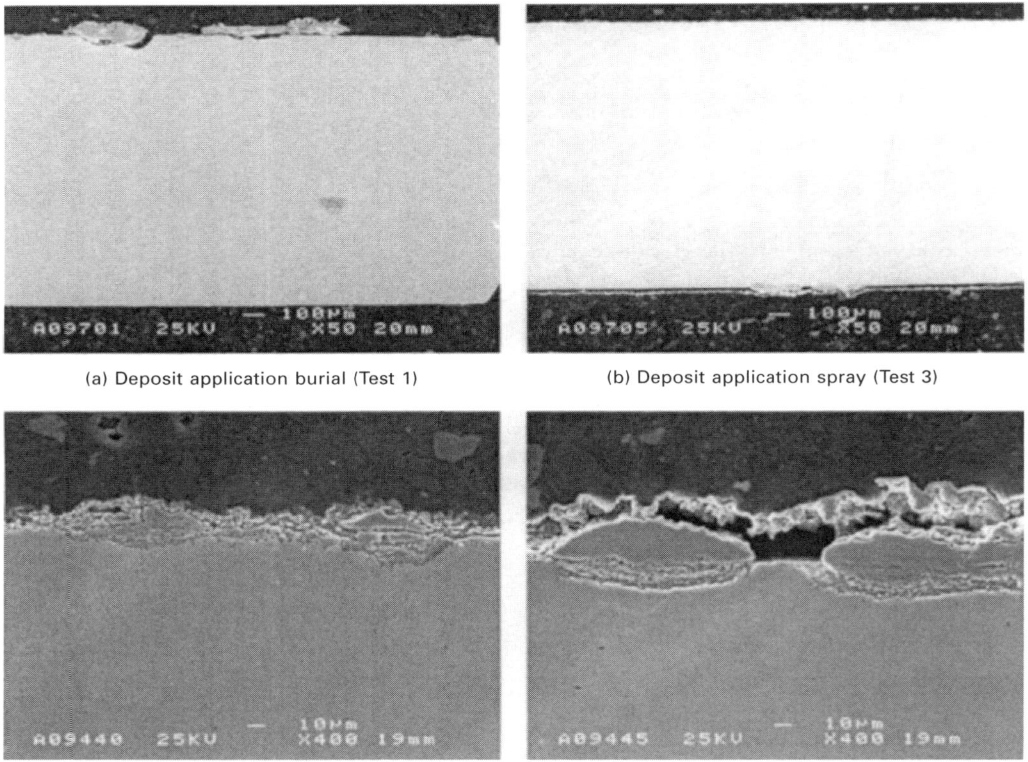

(a) Deposit application burial (Test 1)

(b) Deposit application spray (Test 3)

(c) Deposit application spray, renewal after every 8 cycles (Test 4)

(d) Deposit application spray, renewal after every 4 cycles (Test 3)

11.1 Scale morphologies of specimens of alloy P91 tested at 650 °C. Hot dwell time 20 h, total hot dwell time 300 h. Cross-sectional SEM images.

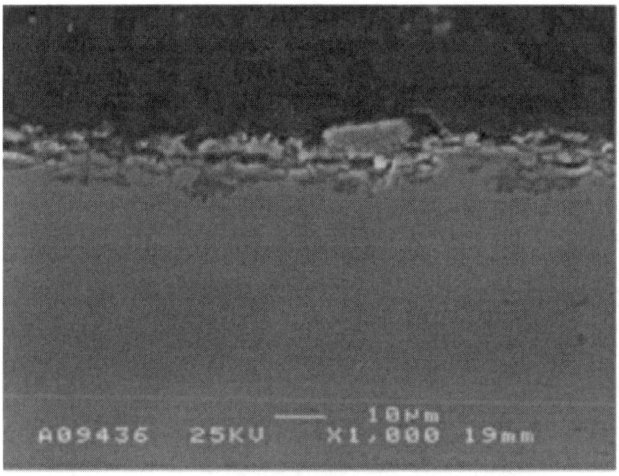

(a) Deposit application spray, renewal every 14 cycles (Test 5)

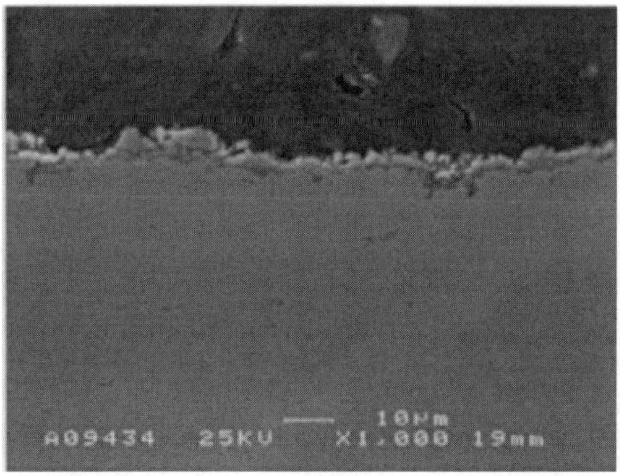

(b) Deposit application dipping, renewal every 14 cycles (Test 6)

11.2 Scale morphologies of alloy AISI441 tested at 850 °C. Hot dwell time 8 h, total hot dwell time 224 h. Cross-sectional SEM images.

recommended as the deposit application method to be applied in cyclic oxidation testing.

Under a humidified atmosphere (Test 7), minute negative gross and net mass changes were observed for salt-coated, but not for plain, AISI 441 specimens. The reason for this behaviour is not obvious and should be studied in further work. However, salt-coated specimens showed cracked, obviously non-protective corrosion scales and some internal oxidation attack (Fig. 11.3).

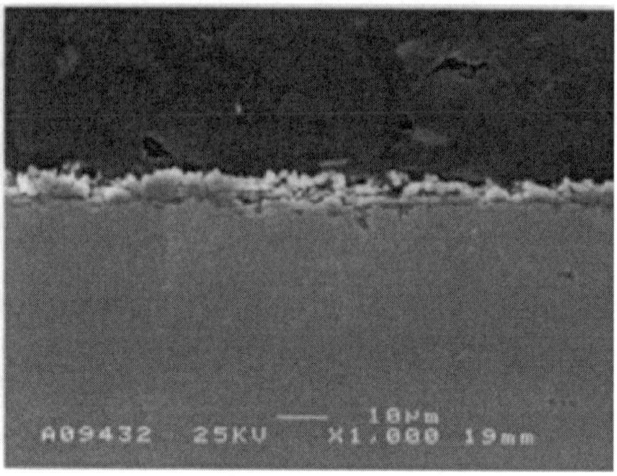

(a) Deposit application method spray, renewal after every 14 cycles

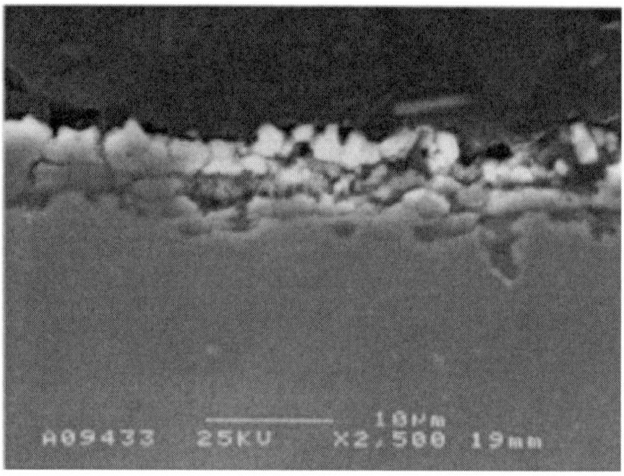

(b) Cross-section shown in (a) examined using a higher magnification of SEM

11.3 Scale morphology of AISI 441 specimen exposed under humidified atmosphere at 850 °C. Cross-sectional SEM images.

11.4 Development of a draft Code of Practice for thermal cycling oxidation testing under deposit conditions

The draft Code of Practice for cyclic oxidation testing under deposit conditions was prepared within the lines of the main code. However, several complications as compared to the main code remain unresolved. The deposit testing guidelines are not included in the main text but as an enclosure of the main draft Code of Practice.

Single salt composition (K_2SO_4) applied by spraying may be used in cyclic oxidation testing at 650 °C. However, this salt composition is not appropriate for use in burner rig cyclic oxidation testing. Instead, a salt mixture of $Na_2SO_4 : K_2SO_4$ corresponding to the eutectic composition with a melting point of about 830 °C (0.794 mol% Na_2SO_4 and 0.206 mol% K_2SO_4) shall be applied at a target level of 0.6 mg/cm^2. Dry synthetic air containing 0.5 vol.% SO_2 humidified to contain 2.5 vol.% H_2O is recommended. If difficulties are encountered with the formation of H_2SO_4 in the furnace, the SO_2 (or H_2O) may be omitted.

11.5 Validation of the draft Code of Practice for cyclic oxidation testing under deposit conditions

The materials tested were AISI 441 and P91. The SO_2 component was omitted from the test atmosphere. The furnace arrangement was the same as in WP5B. The practical importance of the cold dwell time in the case of horizontal furnace arrangement was addressed by using 18 h/6 h cycle for the former and 20 h/4 h cycle for the latter material.

11.5.1 Testing procedure

The applied test parameters and detailed description of the spray application of the deposit are given in Table 11.3. The principles of preparing the mixture for deposit application by spraying and its application were as follows: to obtain the deposit composition (0.75 mol% Na_2SO_4/0.25 mol% K_2SO_4) the given mixture in Table 11.4 was made:

Table 11.3 Test parameters applied in the validation tests at VTT

Gas flow rate	6 l/h
Upper dwell temperature	650 and 850 °C
Dwell time at upper temperature	20 and 18 h
Lower dwell temperature	25 °C
Dwell time at lower temperature	4 and 6
Heating and cooling rate	≈1 cm/min
Gas environment	Synthetic air
Humidity	2.5%
Deposit amount	≈ 0.6 mg/cm^2

Table 11.4 Deposit composition

	g/mol	mol%	mol	g
Na_2SO_4	142.05	0.75	10	10.65375
K_2SO_4	174.27	0.25	10	4.35675
H_2O	18	99	10	178.2
Total		100	10	193.2105

The total amount of moles chosen was 10 mol, therefore the total weight of the solution was 193.2105 g. The salts were first mixed together (10.653 75 g of Na_2SO_4 and 4.356 75 g of K_2SO_4) and after that the water (178.2 g) was added; the salts were left to dissolve for one night before spraying. The amount of solution produced was enough for the whole test to ensure a uniform deposit composition throughout the exposure tests. The equipment used for the spraying system was a small paint sprayer (airbrush) and carrier gas (compressed air) for salt solution. The specimens with crucible were weighed at room temperature prior to the deposit application. The specimens were preheated up to 180 °C and the temperature was left to become even (>10 min). The specimen was sprayed with deposit flux using the injector fixed on horizontal position on the rack 16 cm apart from the specimen. Prior to deposit spraying, test spraying was done to ensure uniformity of spray and the correct, even coverage area. The specimen (an axially symmetrical part could be rotated during the spraying to obtain an even deposit layer) was sprayed with a fixed amount of deposit for a fixed time. After spraying, the specimen was cooled in the desiccator and when the deposit layer was dry the specimen was weighed at room temperature to check the amount of applied deposit. The spraying was repeated until the desired amount of deposit had been obtained.

11.5.2 Test results

No particular technical difficulties were encountered even in the case of the 20 h/4 h cycle. This was due to the deposit reapplication programme (after 8, 12, 16 cycles), which allowed, within the normal weekly working hours, the manual moving of the sample holder supported by alumna rods. The samples before and after the exposure tests are shown in Figs 11.4 and 11.5.

No spallation was observed to occur, irrespective of the material tested. However, different behaviours of measured net and gross mass changes were observed. In the case of the material P91 the gross and net mass changes measured for each sample or in average were identical within the weighing accuracy, Fig. 11.6. Apparent protective-type behaviour was observed up to 120 h of hot dwell time, i.e. until the first deposit reapplication. After that increasing weight gain was observed. The effect of the mass of the deposit applied (in total about 2.5 mg/cm^2) must be considered separately.

In the case of the alloy AISI 441 no protective behaviour could be shown to exist under the test conditions applied (850 °C, wet atmosphere). After an apparent initial weight gain, negative mass change was measured until the first deposit reapplication was made. On average, a clear and approximately linear negative net mass change was calculated to occur Fig. 11.7. A similar negative mass change was observed for the AISI 441 specimens coated with

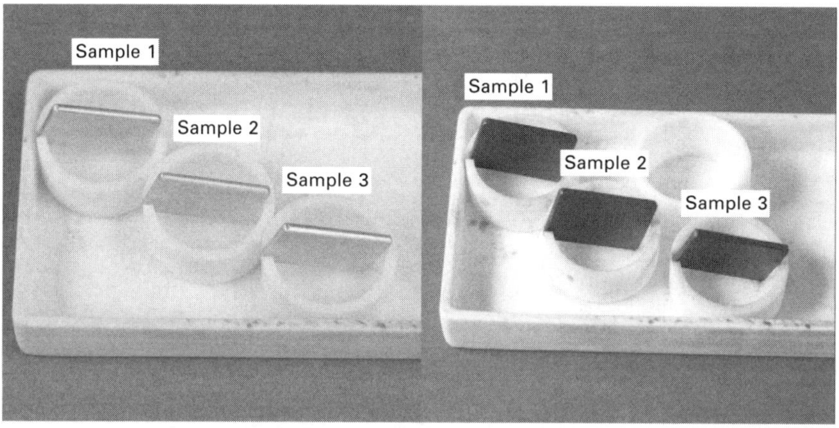

11.4 The samples AISI 441 before (a) and after (b) the furnace exposure.

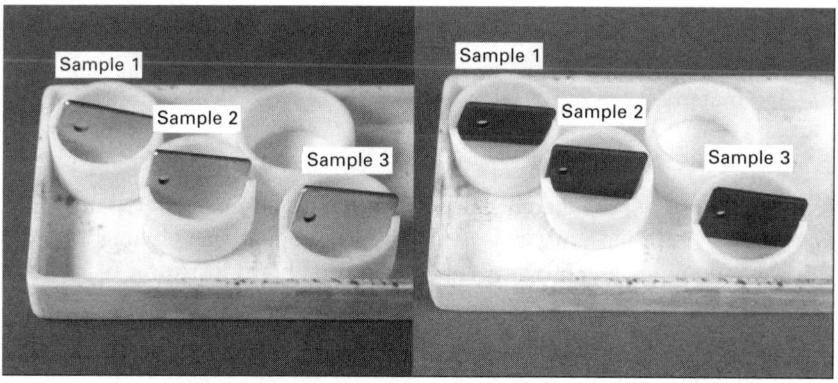

11.5 The samples P91 before (a) and after (b) the furnace exposure.

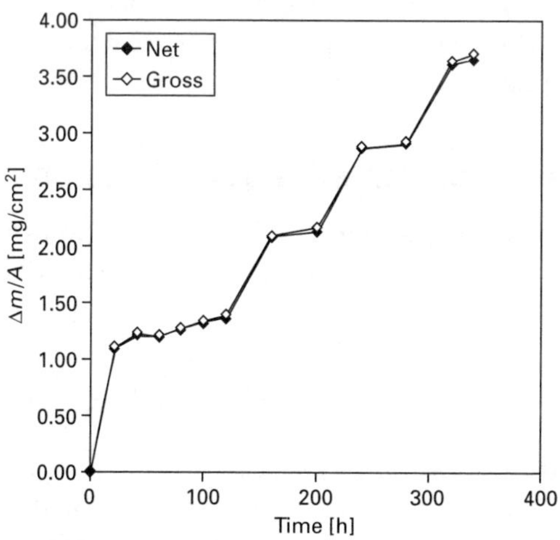

11.6 Average net and gross mass change calculated for the P91 material.

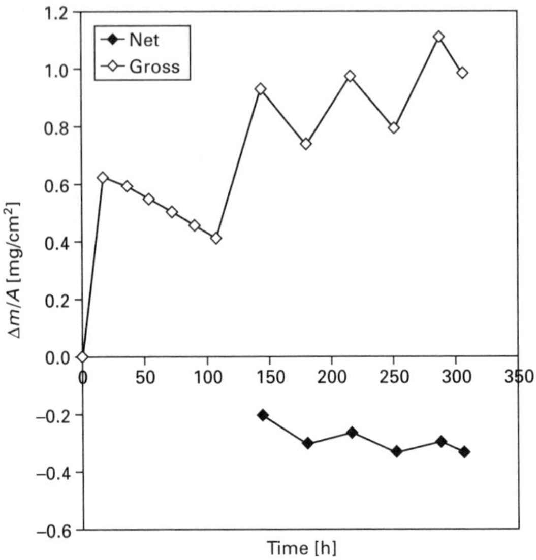

11.7 Average gross and net mass change calculated for the AISI 441 material.

K_2SO_4 salt, but not for the uncoated specimen tested under wet atmosphere in WP5B2. From the gross mass change behaviour it was obvious that the salt applied (single and mixture salt composition both) lost some of its mass due to decomposition/evaporation. However, the negative gross mass change measure after 306 h hot dwell time may be at least in part due to volatilisation of the applied deposit components. It is, however, proposed that volatile corrosion products of the base material contribute to the negative net mass change. This question will be discussed in some detail later, based on the scanning electron microscopy/energy dispersive spectrometer (SEM/EDS) studies of the corrosion scales.

11.6 Post-exposure characterisation of the samples

The samples were analysed with SEM/EDS after the exposure from the surface. Light rinsing with ethanol was performed to the samples before the analysis, as well as coating with a thin carbon layer. Specimens were studied with scanning electron microscope Jeol JSM-6400, and analysed with an energy dispersive X-ray analyser (EDS) PGT Spirit, which can detect carbon and heavier elements.

Various surface morphology details of the scales on the P91 samples are shown in Fig. 11.8. The scales were not detached by brushing and individual protruding grains seem to be firmly attached to the surface. SEM/EDS analyses

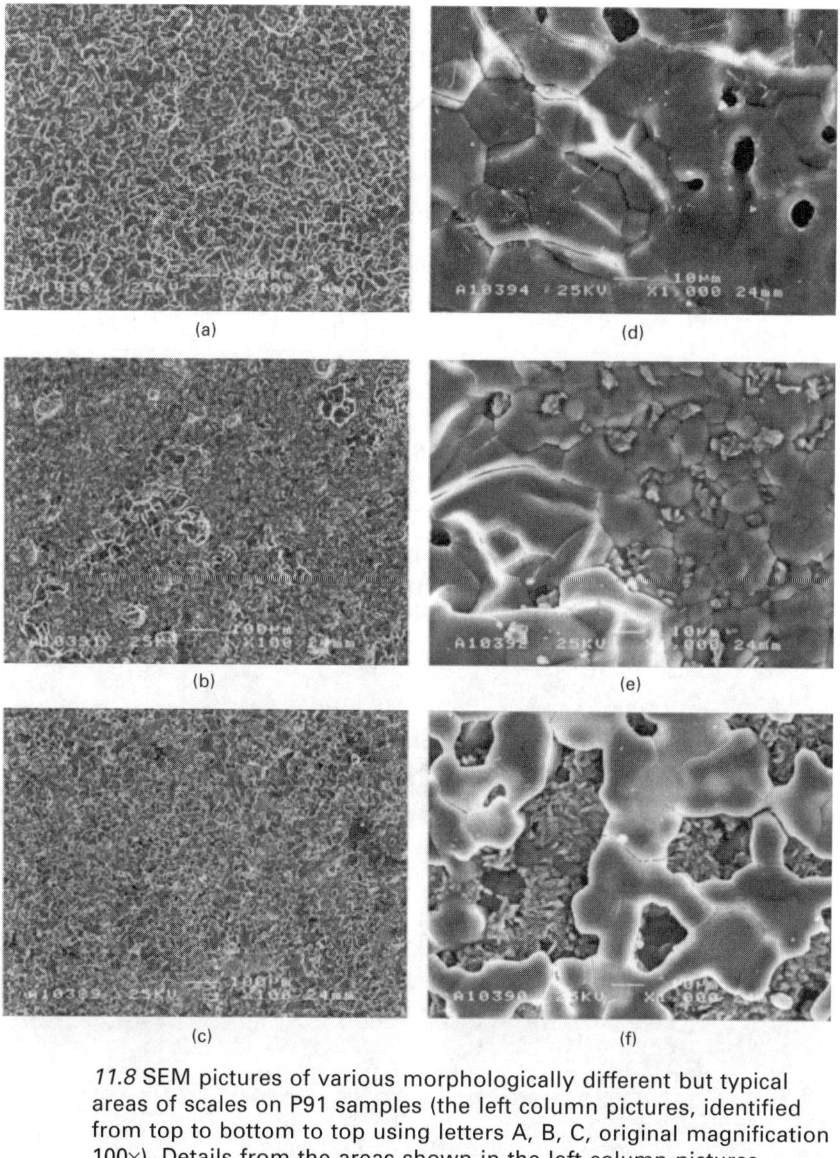

11.8 SEM pictures of various morphologically different but typical areas of scales on P91 samples (the left column pictures, identified from top to bottom to top using letters A, B, C, original magnification 100×). Details from the areas shown in the left column pictures viewed (the right column pictures, original magnification 1000×).

of various surface areas and their details (see Table 11.5) revealed that the scales were composed of components of the applied deposit and of complex corrosion products (e.g. mixed oxides and sulphates). The scales appear permeable to the oxidising atmosphere. It is obvious that the complex scale may absorb the stresses created during cooling and heating so that the scales do not easily spall. However, the scales do not provide protection against further deposit corrosion during repeated exposures at the test temperature.

Significant evaporation of the applied deposit (or corrosion products formed) seems not to occur; in accordance with that, the vapour pressure of both mixture components is low at 650 °C.

Various surface morphology details of the thin scales found on the surfaces of the AISI 441 samples are shown in Fig. 11.9. SEM/EDS analyses of

Table 11.5 Example quantitative analyses determined using from various areas (Types A–C) of the outer scale surface of the material P91. Magnification used 100× and 1000×. Weight fractions given are indicative only

	Type A		Type B		Type C	
	100×	1000×	100×	1000×	100×	1000×
Cr	6.5	6.5		0.5		0.5
K	29.9	29.8	15.6	17.9	19.3	23.7
S	34.2	33.6	29.9	26.7	34.8	33.5
Na	28.4	28.8	32.4	30.6	31.6	32.2
Fe	1.0	1.3	20.8	22.8	12.6	8.5
Mn			1.4	1.5	1.7	1.6
Cu						

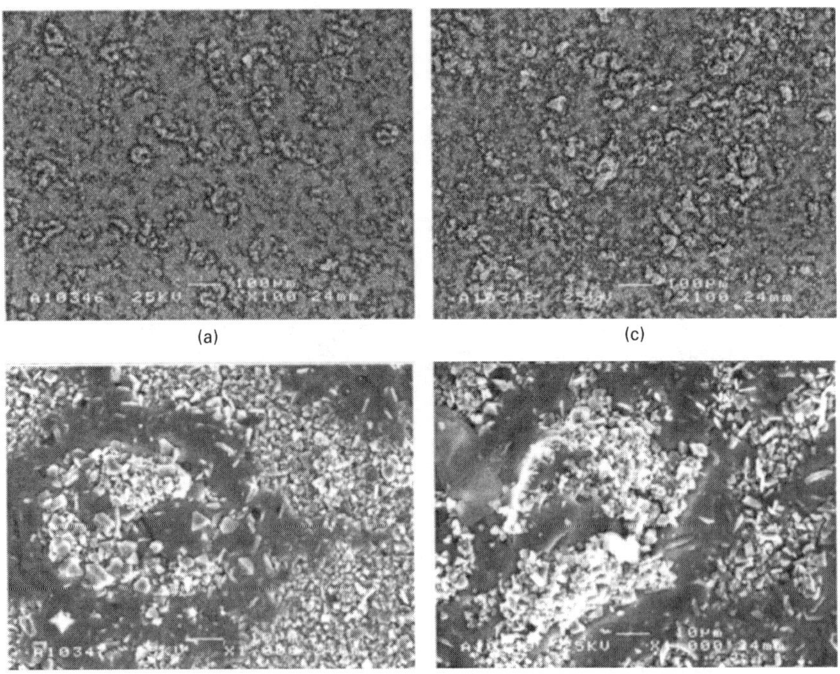

11.9 SEM images of scales on AISI 441 material samples. Left column sample no. 2, right column sample no. 3. Original magnification: top row 100×, bottom row 1000×. SEM pictures.

various surface areas and some point analyses of individual phases are given in Table 11.6. Locations of point analyses are indicated in Fig. 11.10.

The results of the area analyses performed indicate that the deposit might become impoverished, as compared with the content of sodium, in respect of the content of potassium and to a lesser extent of that of sulphur. Selective evaporation and minute mass loss of the applied deposit may occur during the hot dwell time.

Table 11.6 Quantitative EDS analyses of the outer scale surface area of the sample 3 shown in Fig. 11.9. Magnification used 100×, 1000× (area analyses) 5000× (point analyses from 1 to 4 represents the various phases shown in Fig. 11.10). Weight fractions given are indicative only

wt% of	100×	1000×	5000× (1)	5000× (2)	5000× (3)	5000× (4)
Cr	56.8	60.26	42.7	29.5	31.0	19.1
Si	0.9	1.95	0.5	1.4	0.9	0.3
O			21.0	39.2	26.5	32.1
S	8.6	5.5		1.8	2.1	11.7
K	3.3	1.7		0.6	0.5	5.0
Ti	0.9	1.4	0.4	0.8	0.3	0.1
Cu	0.6	0.8	2.3	0.7	1.1	
Ca	0.6	0.5		6.3	0.5	0.1
Na	20.4	15.6		5.0	9.3	20.8
Nb	2.8	2.9				
Mn		9.5	29.9	12.6	17.6	1.1
Fe	4.3		3.2	1.5		1.2
Mg				0.7	0.0	
C					10.3	8.7

11.10 SEM picture showing the locations of the point analyses referred to in Table 11.6

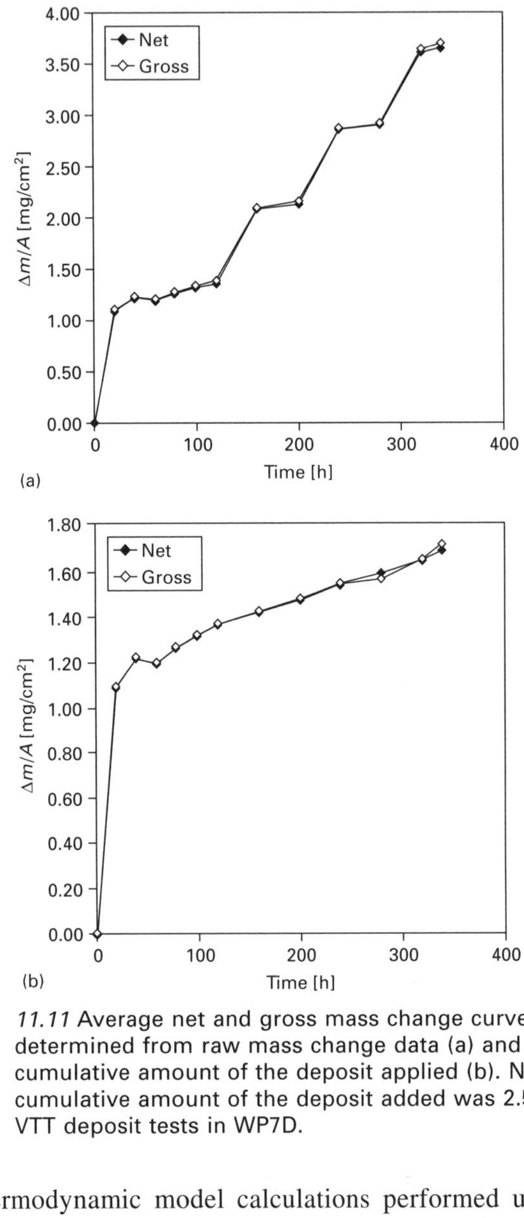

11.11 Average net and gross mass change curves for P91 specimens determined from raw mass change data (a) and after subtracting the cumulative amount of the deposit applied (b). Note that the total cumulative amount of the deposit added was 2.5 mg/cm^2. Data from VTT deposit tests in WP7D.

Thermodynamic model calculations performed using the FACTSage database, indicate that there might indeed occur some volatilisation of sulphur as oxides and some of alkalis as hydroxides, but the total vapour pressure of deposits as applied is at theoretic equilibrium only in the range of 10^{-6} atm at 850 °C. However, when a small amount of chromium was added to the thermodynamic model system, liquid sodium chromate ($Na_2CrO_{4(l)}$) was formed, indicating that liquid phase fluxing of the chromium oxide scale may occur during the test. The thermodynamic database used did not include various gaseous complex species possibly contributing to the metal loss. It is, however, possible to make a qualitative conclusion that the metallic samples have suffered significant metal loss by evaporation of volatile corrosion

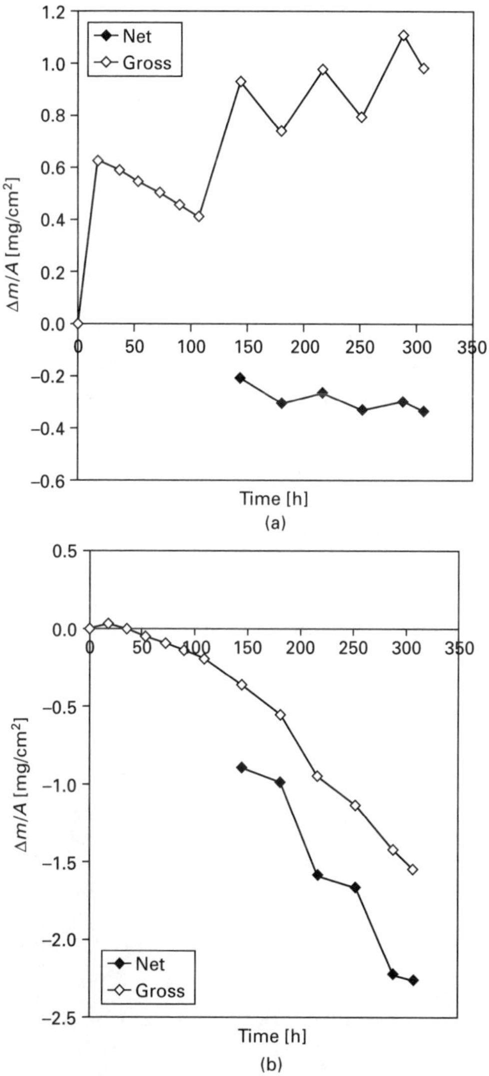

11.12 Average net and gross mass change curves for AISI 441 specimens determined from raw mass change data (a) and after subtracting the cumulative amount of the deposit applied (b). The appearance of the curves implies that significant evaporation loss during the test occurs. Note that the total cumulative amount of the deposit added is about 2.5 mg/cm^2. Data from VTT deposit tests in WP7D.

product species. More detailed discussions of this problem are beyond the scope of this project. Without detailed additional chemical analyses to determine the actual mass balances, it is not possible to assess the actual gross and net mass change behaviour of the AISI 441 samples.

11.7 Conclusions

The work performed indicates that the module developed for deposit type of complex corrosion testing is widely applicable. A more stable salt mixture than the one used previously in TESTCORR was found to be satisfactory. The analysis of the test results in such cases where significant deposit and/or volatile corrosion product evaporation to the gaseous environment occur, is not straightforward, and may in extreme cases lead to the mass change data measured is factually indeterminate. However, the particular problems related to applying test atmosphere variants omitting SO_2 or H_2O are understood better now than before. As already emphasised at the beginning of this project, the long-term stability of the deposit under non-equilibrium gas atmosphere conditions is the principal problem to be addressed.

For the sake of clarity, the cumulative amount of the deposit applied should be subtracted from the mass change measured in the deposit testing. In the optimal case, where there are no significant evaporation losses, the corrosion kinetics may be deduced from the mass change curves as defined in the main Code of Practice (see Fig. 11.11). In the most adverse cases, significant evaporation losses during the test occur (see Fig. 11.12), making the evaluation of corrosion kinetics arbitrary when based on the mass change curves only.

Part III

Code of Practice

12
Validation testing of the Code of Practice and statistical analysis of experimental results

J. R. NICHOLLS, Cranfield University, UK,
S. COLEMAN, Newcastle University, UK and
M. MALESSA and M. SCHÜTZE,
DECHEMA e.V., Germany

12.1 Introduction

A validation test matrix was designed with due statistical consideration to permit comparison between laboratories undertaking the evaluation and to permit evaluation of the guidelines application to a range of materials under test. This was a major deliverable for the second part of the COTEST project. This work package was coordinated by Cranfield University and involved 22 laboratories, as either partners or subcontractors.

12.2 Validation test matrix

Four 'round robin' tests were planned covering:

Sub-Task 7A Long dwell times
Sub-Task 7B Short dwell times
Sub-Task 7C Ultra-short times
Sub-Task 7D Complex environments

The plan for these tests was developed to allow comparison of results for:

- different alloys in the same institution;
- different dwell times for the same alloy and for the same institution; and
- in some cases, different environments.

The test design aimed at providing maximum opportunity for comparing the effects of laboratory, alloy and testing regime while limiting the amount of testing required by each laboratory. The conceptual aim was to achieve that:

- each laboratory will test three materials;
- each material will be tested in three laboratories;
- each material will be tested under reference conditions; and
- triplicate specimens will be included in all tests.

Materials were produced by project partners to meet the needs of this plan and specimens were manufactured. Samples were characterised at Centro Elettrotecnico Sperimentale Italiano (CESI) and then distributed to the laboratories for testing.

The plan was modified slightly just prior to the commencement of the validation testing to accommodate practical constraints between the collaborating laboratories. The resulting design was balanced as far as possible. This validation study concerns the first part of the validation testing for cyclic oxidation using long and short dwell cycles, with natural cooling in an air environment.

The matrices are given in the Tables 12.1–12.6. Ticked items are tests that were completed to plan. Where specimens were supplied, but of the wrong materials (in error), this has been noted.

Table 12.1 Validation test matrix for long dwell testing

Material Test temperature (°C)	P91 650	AISI 441 850	Alloy 800H 1000	CM 247 1150	Kanthal A1
CESI	✓			✓	
Cranfield		✓			CM 247 samples
DECHEMA			✓		
DLR				✓	
EC-JRC-IAM	Not heat-treated	✓			
FZJ			✓		✓
NPL		✓	✓		
University Siegen				✓	
SIMR	✓				✓

Table 12.2 Validation test matrix for short dwell testing

Material Test temperature (°C)	P91 650	AISI 441 850	Alloy 800H 1000	CM 247 1150	Kanthal A1
Arcelor		✓	✓		
CESI	✓				
Cranfield		✓			
DECHEMA		✓	✓		✓
DLR				✓	
FZJ	✓		✓		
University Liverpool					✓
MTU				✓	
NPL	✓				
University Siegen				✓	
SIMR					✓

Table 12.3 Validation test matrix for ultra-short dwell testing

Material Test temperature (°C)	P91 650	AISI 441 850	Alloy 800H 1000	CM 247 1150	Kanthal A1
Cranfield			✓		✓

Table 12.4 Validation test matrix for testing in complex corrosive gases

Material Test temperature (°C)	P91 650	AISI 441 850	Alloy 800H 1000	CM 247 1150	Kanthal A1
Alstom				✓	
CESI	✓	✓	✓		
UCM	✓	✓	✓		

Table 12.5 Validation test matrix for deposit corrosion testing

Material Test temperature (°C)	P91 650	AISI 441 850	Alloy 800H 1000	CM 247	Kanthal A1
UCM	✓	✓	✓		
VTT	✓	✓			

Table 12.6 Validation test matrix for burner rig tests

Material Test temperature (°C)	P91 650	AISI 441 850	Alloy 800H 1000	CM 247	Kanthal A1
MTU (high velocity)				✓	
NPL (low velocity)	✓	✓	✓		

COTEST partners and subcontractors carried out the tests according to the draft Code of Practice developed in Work Package 6 and submitted their results using the database designed in Work Package 2. The spreadsheets provided automatic plots and the experimenter calculated the summary statistics of the net change profile in accordance with the Code of Practice. As the results of the various tests from Work Package 7 became available, the COTEST coordinator presented the net mass change curves on the COTEST website. The net mass change curves for the different tests are given in Figs 12.1–12.15. Sufficient data were returned from this validation study to permit the statistical analysis of the data both between laboratories for selected alloys and between test types (long and short dwell times, etc).

While in most cases a good agreement between the net mass change curves of identical alloys from different test facilities was found, in some

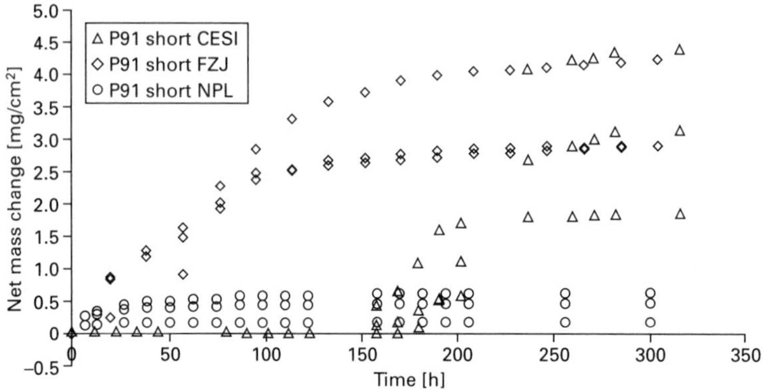

12.1 Net mass change of P91 during short dwell thermal cycling oxidation (650 °C, 1 h hot, 0.25 h cold, wet air, heat-treated specimens).

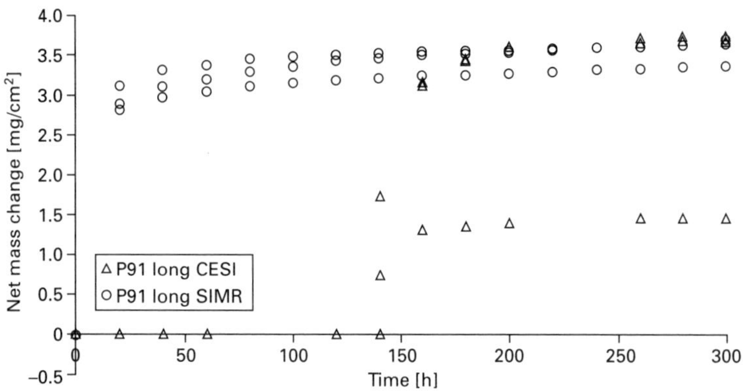

12.2 Net mass change of P91 during long dwell thermal cycling oxidation (650 °C, 20 h hot, 4 h cold, wet air, heat-treated specimens).

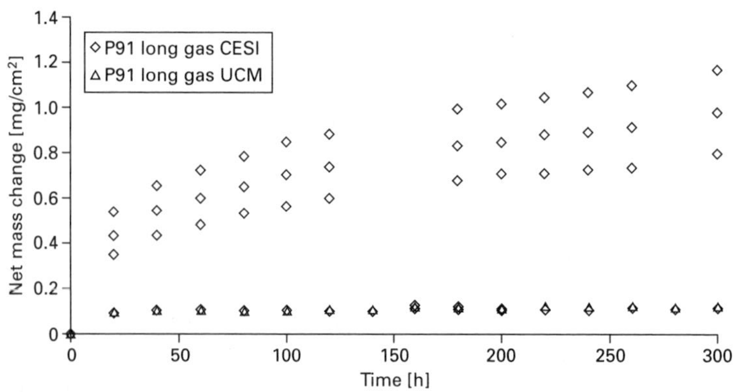

12.3 Net mass change of P91 during thermal cycling oxidation in SO_2-containing atmosphere (650 °C, 20 h hot, 4 h cold, dry air/5000 ppm SO_2, heat-treated specimens).

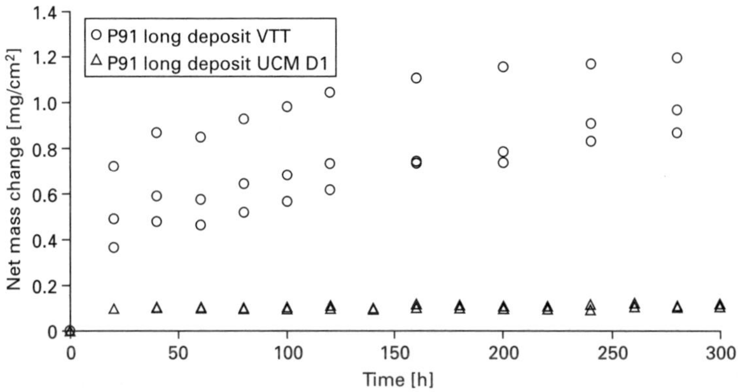

12.4 Net mass change of P91 during thermal cycling oxidation under deposit application (650 °C, 18 h hot, 6 h cold, dry air/deposits, heat-treated specimens).

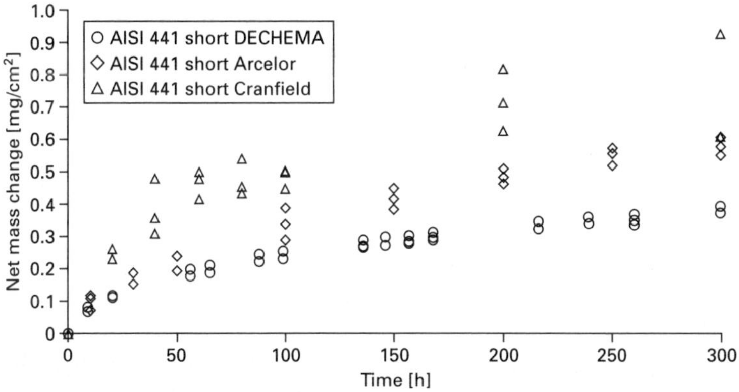

12.5 Net mass change of AISI 441 during short dwell thermal cycling oxidation (850 °C, 1 h hot, 0.25 h cold, dry air).

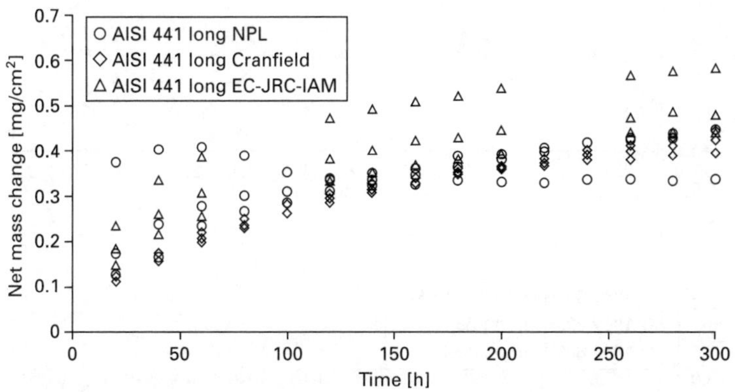

12.6 Net mass change of AISI 441 during long dwell thermal cycling oxidation (850 °C, 20 h hot, 4 h cold, dry air).

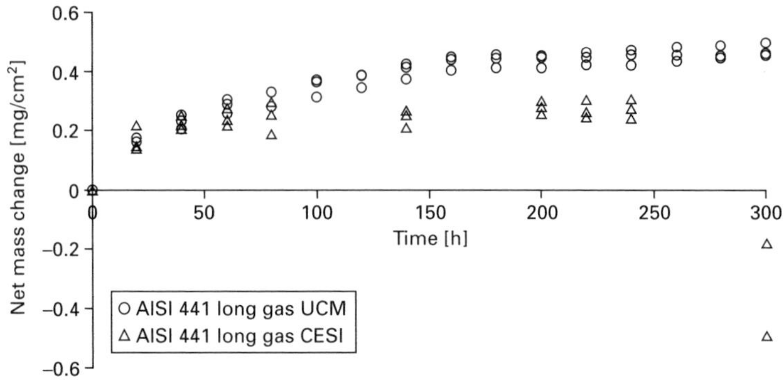

12.7 Net mass change of AISI 441 during long dwell thermal cycling oxidation in SO_2-containing atmosphere (850 °C, 1 h hot, 0.25 h cold, dry air/5000 ppm SO_2).

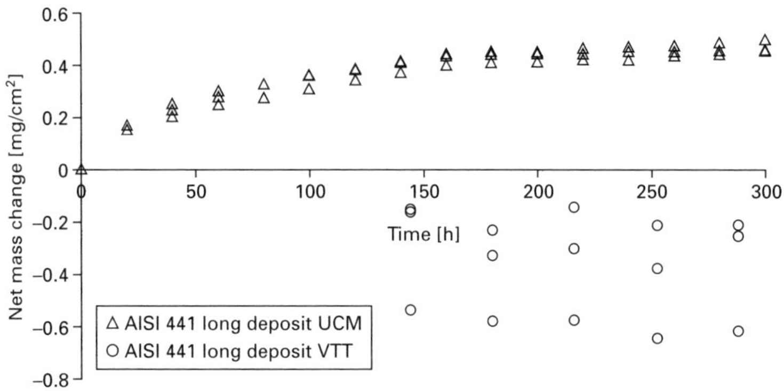

12.8 Net mass change of AISI 441 during long dwell thermal cycling oxidation under deposit application (850 °C, 18 h hot, 6 h cold, dry air/deposits).

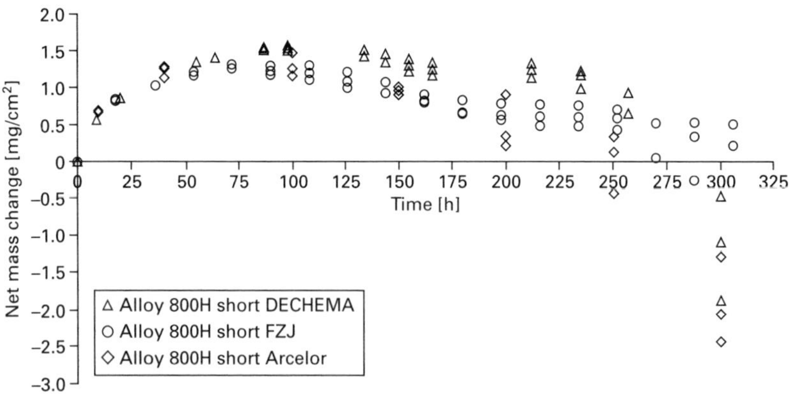

12.9 Net mass change of Alloy 800H during short dwell thermal cycling oxidation (1000 °C, 1 h hot, 0.25 h cold, dry air).

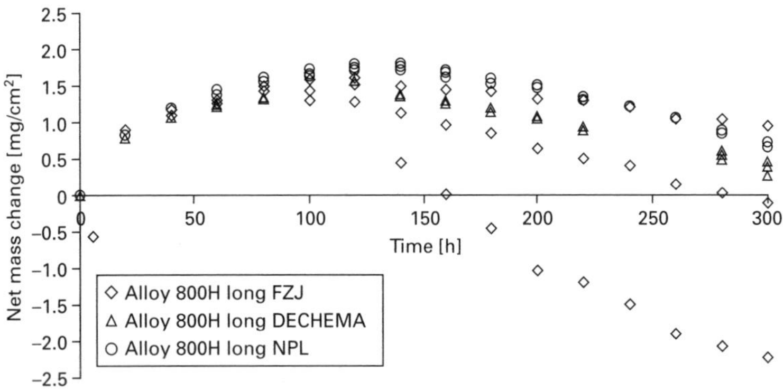

12.10 Net mass change of Alloy 800H during long dwell thermal cycling oxidation (1000 °C, 20 h hot, 4 h cold, dry air).

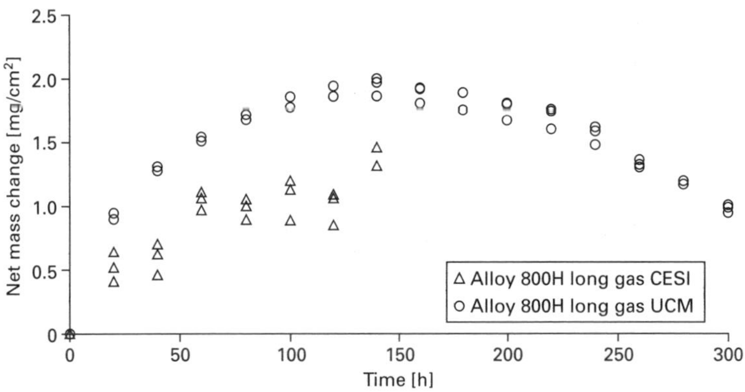

12.11 Net mass change of Alloy 800H during long dwell thermal cycling oxidation in SO_2-containing atmosphere (1000 °C, 20 h hot, 4 h cold, dry air/5000 ppm SO_2).

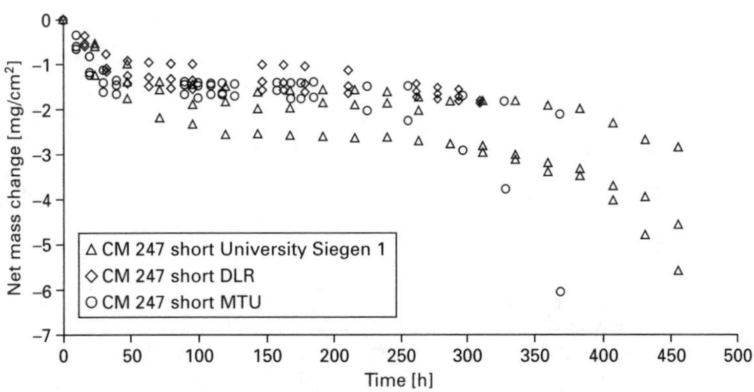

12.12 Net mass change of CM 247 during short dwell thermal cycling oxidation (1150 °C, 1 h hot, 0.25 h cold, dry air).

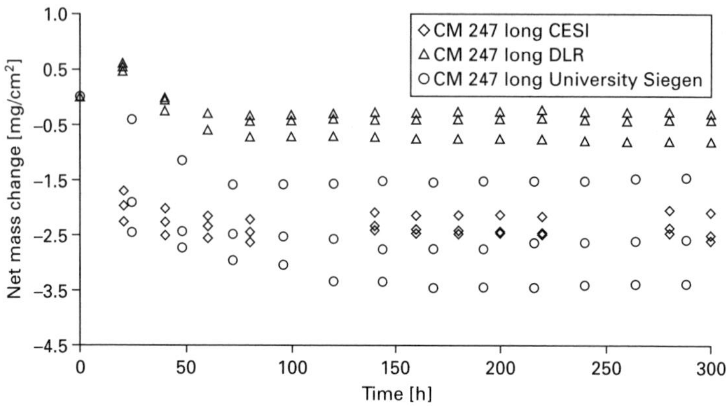

12.13 Net mass change of CM 247 during long dwell thermal cycling oxidation (1150 °C, 20 h hot, 4 h cold, dry air).

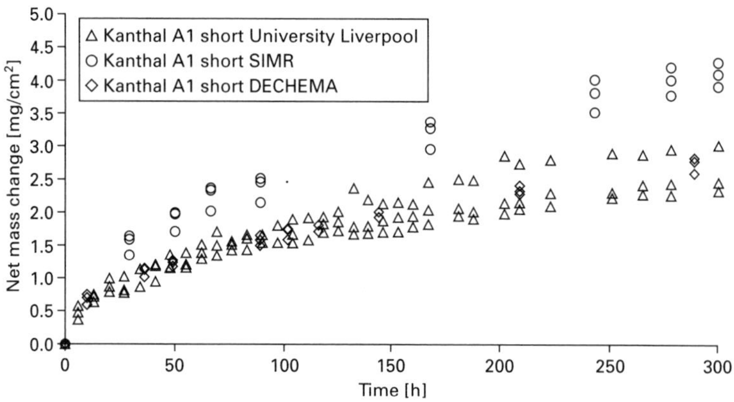

12.14 Net mass change of Kanthal A1 during short dwell thermal cycling oxidation (1250 °C, 1 h hot, 0.25 h cold, dry air).

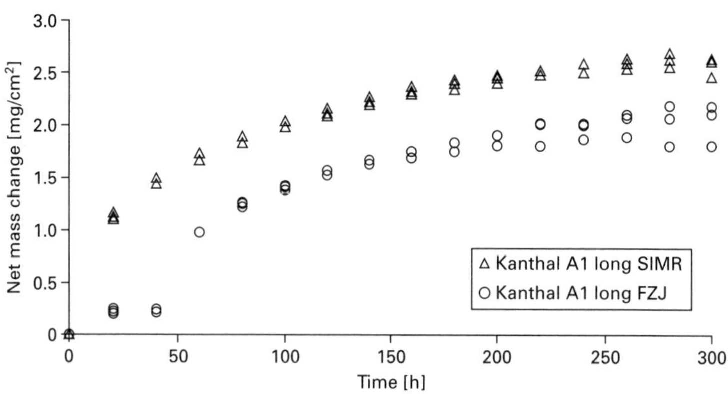

12.15 Net mass change of Kanthal A1 during short dwell thermal cycling oxidation (1250 °C, 1 h hot, 0.25 h cold, dry air).

cases deviations were obvious. For the alloy P91 the origin of these deviations is attributed to the sensitivity of oxidation behaviour of the alloy after being subjected to the heat treatment process. The heat treatment had to be performed in several batches which were carried out in one laboratory under controlled conditions. However, batch-to-batch variations obviously occurred, possibly because of small differences in the temperature profile of the heat treatment profile. The draft Code of Practice applied in the validation testing is therefore powerful enough to distinguish between the different batches.

In general the results from long and short dwell testing exhibited good agreement, while test results in complex corrosion testing (complex corrosive atmospheres or testing under deposit application) were found to show more variations. The origin for these variations especially under deposit conditions is attributed to the lab-to-lab variations of test environments, i.e. deposit application procedure.

Another source for differences in oxidation behaviour is the temperature profile of the thermal cycle. While the heating rate does not influence the oxidation behaviour significantly (see Chapter 5), the cooling rate might influence the oxidation behaviour by the impact on spallation due to stresses caused by thermal expansion. The temperature profiles of the thermal cycles were recorded prior to the exposure of the specimens and are shown in Figs 12.16 to 12.24.

12.3 Analysis of experimental data

The summary statistics were entered into a Minitab project file and Excel by Industrial Statistics Research (ISRU) for analysis. Each validation experiment was carried out on three specimens. The statistical analysis makes use of the mean and log standard deviation of the three results. Values of n and k_n were available for AISI 441, Alloy 800H, Kanthal A1 and some P91 but not for

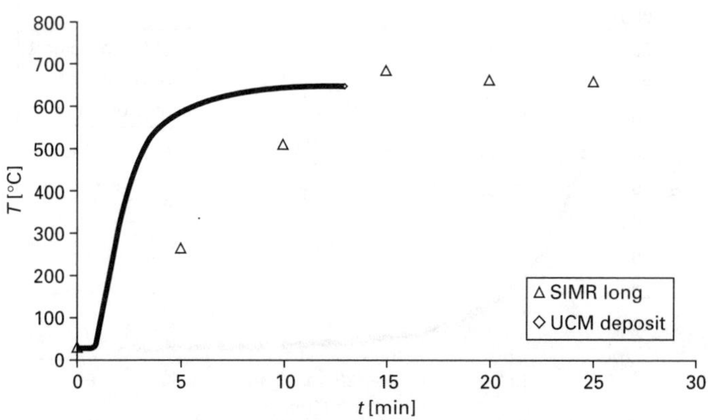

12.16 Temperature profiles of heating period to 650 °C.

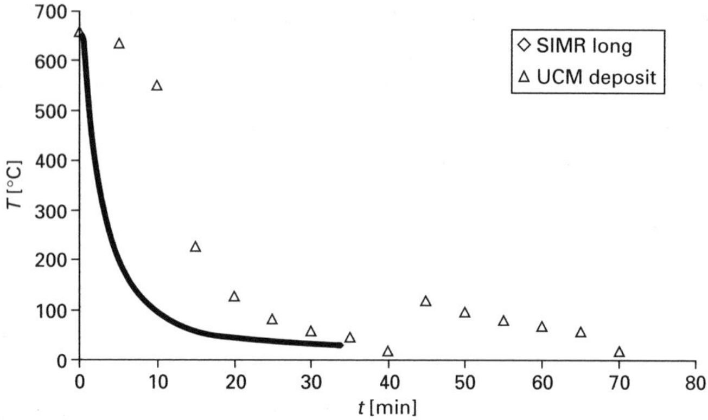

12.17 Temperature profiles of cooling period from 650 °C.

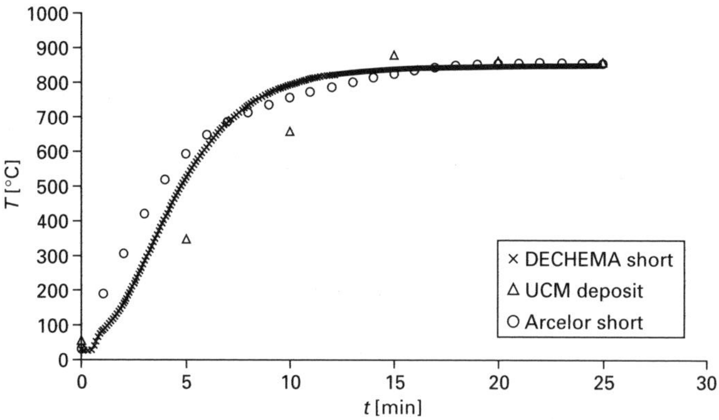

12.18 Temperature profiles of heating period to 850 °C.

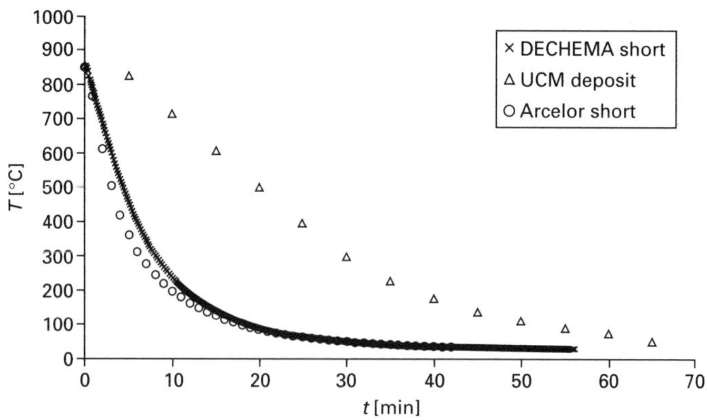

12.19 Temperature profiles of cooling period from 850 °C.

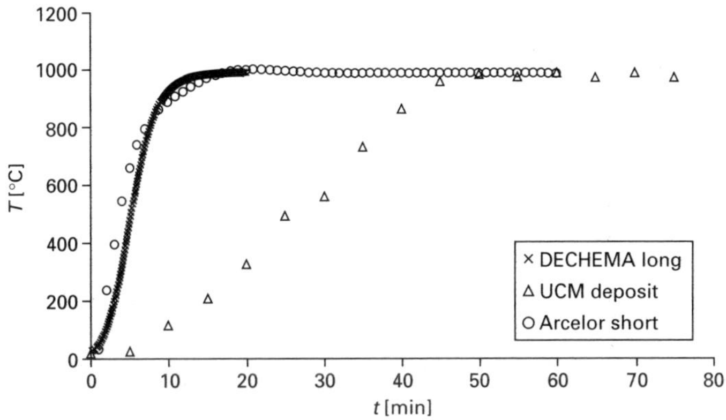

12.20 Temperature profiles of heating period to 1000 °C.

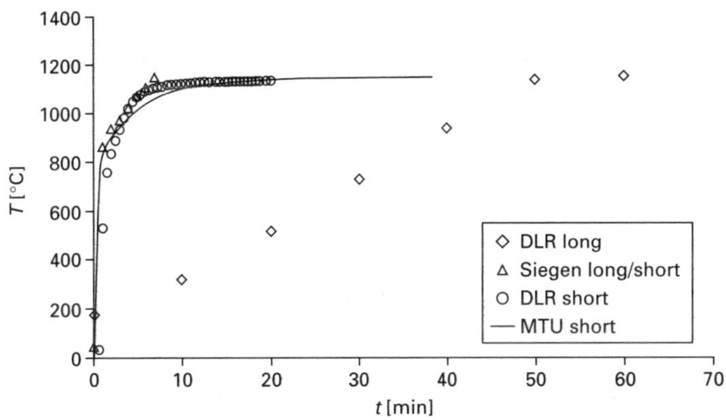

12.21 Temperature profiles of heating period to 1150 °C.

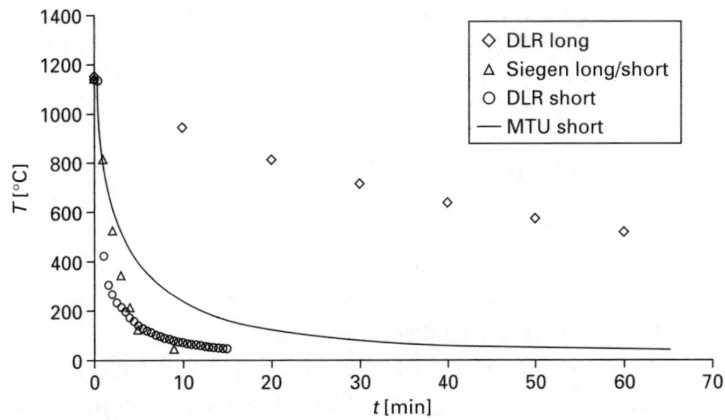

12.22 Temperature profiles of cooling period from 1150 °C.

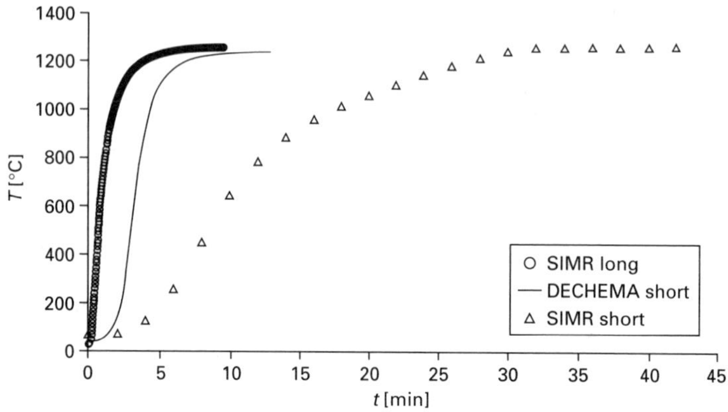

12.23 Temperature profiles of heating period to 1250 °C.

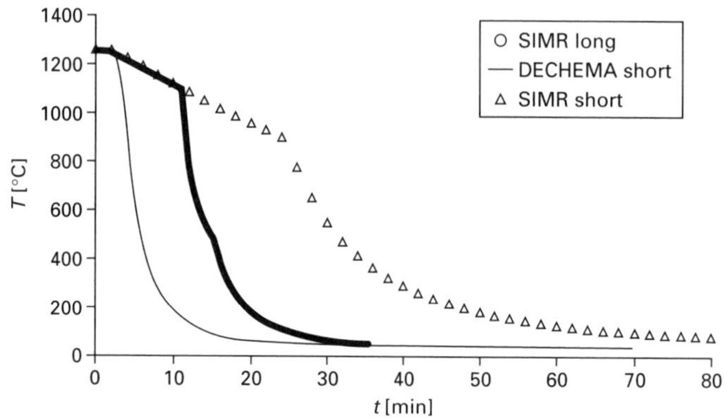

12.24 Temperature profiles of cooling period from 1250 °C.

any CM 247 as the mass change curves were of a totally different shape. Protective oxide growth time $t_{\text{protective}}$ prior to spallation and mass change $\Delta m_{\text{net}}(t_{300} - t_{\text{protective}})$ were also available for AISI 441, Alloy 800H, Kanthal A1 and some of the P91 experiments. In some cases the 300 h testing time did not lead to spallation. The data are shown in Table 12.7.

12.4 Graphical analysis of results

The aim of the validation testing was to compare the results for each alloy when tested in different laboratories. The results from the different laboratories are presented by scatterplots for the different observables (n, k_n, $t_{\text{protective}}$, $\Delta m_{\text{net}}(t_{300} - t_{\text{protective}})$) (Figs 12.25–12.28).

In Fig. 12.25 the scatterplot of n vs the hot dwell time is shown. The results are generally close except for P91 short dwell where the results for NPL and CESI are rather different. As mentioned earlier, these differences

Table 12.7 Data used in statistical analysis

Alloy	Lab	Dwell time (h)	n	k_n (10^{-4})	$t_{protective}$ (h)	Δm_{net} ($t_{300} - t_{protective}$)
AISI 441	Cranfield	20	3.39	21	213	0.05
AISI 441	JRC	20	2.61	8	173	0.06
AISI 441	NPL	20	2.00	7	*	*
Alloy 800H	DECHEMA	20	2.67	288	80	0.93
Alloy 800H	FZJ	20	2.68	375	80	1.89
Alloy 800H	NPL	20	2.18	348	80	*
CM 247	CESI	20	*	*	*	*
CM 247	Cranfield	20	*	*	*	*
CM 247	DLR	20	*	*	*	*
CM 247	Univ Siegen	20	*	*	*	*
Kanthal A1	SIMR	20	2.88	740	160	0.25
Kanthal A1	FZJ	20	4.16	677	*	*
P91	CESI	20	0.04	66	160	*
P91	JRC	20	*	*	*	*
P91	SIMR	20	0.07	*	300	0.00
AISI 441	Arcelor	1	1.86	13	*	*
AISI 441	Cranfield	1	2.50	22	200	0.57
AISI 441	DECHEMA	1	2.16	4	300	0.00
Alloy 800H	DECHEMA	1	2.38	326	98	2.67
Alloy 800H	FZJ	1	3.16	317	72	1.19
CM 247	DLR	1	*	*	*	*
CM 247	MTU	1	*	*	*	*
CM 247	Univ Siegen	1	*	*	*	*
Kanthal A1	DECHEMA	1	2.40	358	300	0.00
Kanthal A1	Liverpool	1	2.36	324	300	0.00
Kanthal A1	SIMR	1	2.38	973	168	0.90
P91	CESI	1	0.13	50	237	*
P91	FZJ	1	*	*	*	*
P91	NPL	1	5.24	10	54	*

* no data.

originate from variations between different heat treatment batches. The similarity of results for the other alloys is commensurate for both long and short dwell.

In Fig. 12.26 the log (to the base e) of k_n is plotted versus the hot dwell time. The use of log values makes their distribution more symmetrical to comply with the usual assumptions for analysis of variance and statistical modelling. The results are generally similar for long and short dwell for each alloy when $\log(k_n)$ is considered.

In Fig. 12.27 the results for $t_{protective}$ for long and short hot dwell times are given. The results are generally similar when $t_{protective}$ is considered, although some isolated differences are observed between laboratories, for one or two of the materials. For example, DECHEMA and Cranfield differ in their reporting of $t_{protective}$ for AISI 441. Cranfield researchers reported that they

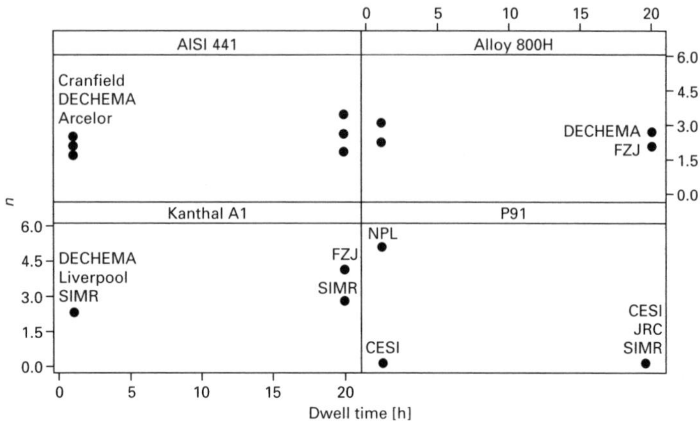

12.25 Scatterplot of n vs hot dwell time.

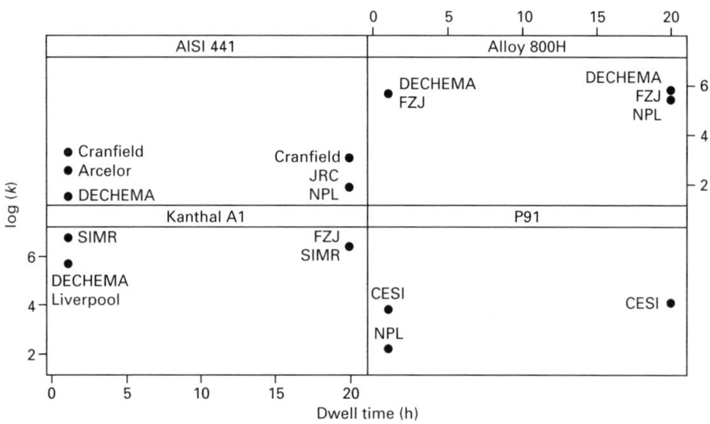

12.26 Scatterplot of log k_n vs hot dwell time.

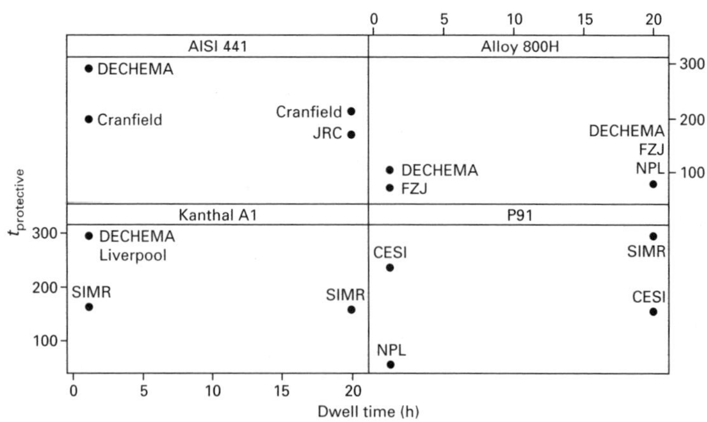

12.27 Scatterplot of $t_{protective}$ vs hot dwell time.

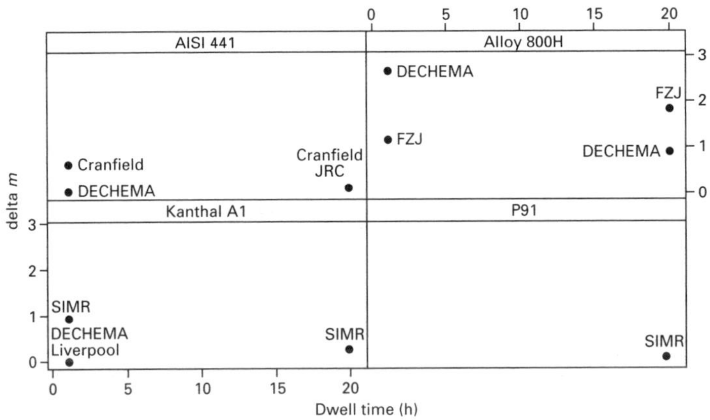

12.28 Scatterplot of $\Delta m_{net}(t_{300\,h} - t_{protective})$ vs hot dwell time.

observed spalling from the onset and thus the lower $t_{protective}$ values: although, based on the calculated regression coefficient, one could confidently report an onset of spall of 200 h, spalled material had not been detected in the crucible after the first cool down cycle.

In Fig. 12.28 the scatterplots for $\Delta m_{net}(t_{300\,h} - t_{protective})$ for long and short dwell times are given for each alloy. Again the results are generally similar when $\Delta m_{net}(t_{300\,h} - t_{protective})$ is considered.

12.5 Statistical analysis

There are 20 usable sets of results for n and 19 for k_n, but for P91 there is obviously more scatter in the experimental results, which is related to the high sensitivity of the oxidation behaviour to the heat treatment process. Therefore some of the analysis has been carried out using only the three alloys AISI 441, Alloy 800H and Kanthal A1. The results are expected to vary for different alloys, laboratories and dwell times. The scatterplot for n shows that most values are between 2 and 3 with the exception of P91 (Fig. 12.29).

The scatterplot for $\log(k_n)$ in Fig. 12.30 shows that most of the variation is due to alloy. This is to be expected as the oxidation rate constant (k_n) depends on the type of scale formed, and whether it is protective or not. Such properties are clearly alloy dependent. The results for long and short dwell times for each alloy are fairly close. The remaining variation is due to laboratory and interactions between these factors.

Analysis of variance (ANOVA) for n shows that the difference between alloys is not significant. Analysis of variance for $\log(k_n)$ shows that the difference between alloys is significant (significance $p = 0.00$).

As expected there is no statistically significant difference between long

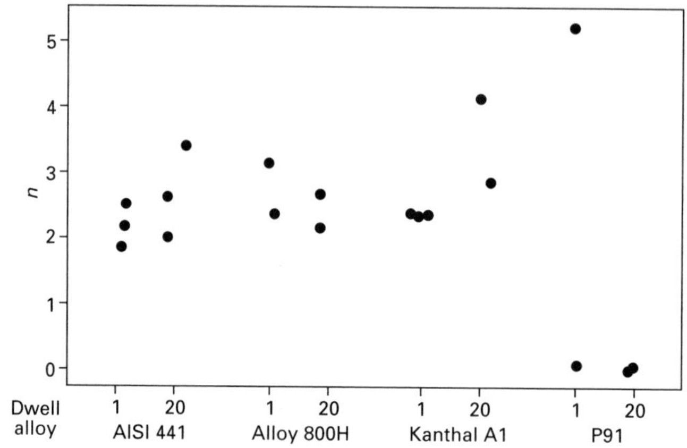

12.29 Scatterplot of *n* vs alloy.

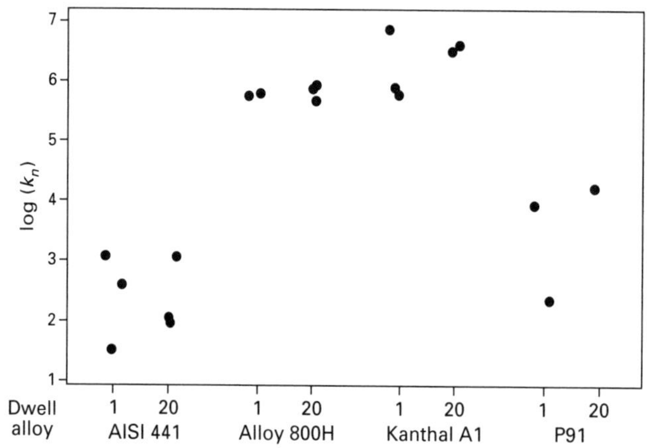

12.30 Scatterplot of log k_n vs alloy.

and short dwell log(k_n) values, as the oxide growth during the heating and the hot dwell period only depends on the oxidation temperature and therefore the oxidation rates are almost identical.

The error term in the analysis of variance (Table 12.8 is made up of the variation between laboratories within alloys). The square root of the error mean square (MS) value is 0.58 and is the estimated standard deviation of laboratories within alloys, given as S in Table 12.8. Further analysis to separate the components of variation shows that the variance component for alloys is 3.91 compared with 0.34 for the laboratories within alloys. The proportion of variation explained by alloys is 92% which equates to the R^2 term in Table 12.8. This is an excellent result showing good correspondence between the results for different laboratories for these alloys.

Analysing the data in terms of the variability of the three sets of results

Table 12.8 Analysis of variance for log (k_n)

Source	Degrees of freedom	Sum of squares	Mean square	F	P
Alloy*	3	55.77	18.59	55.38	0.00
Error	15	5.04	0.34		
Total	18	60.81			

S = 0.58, R^2 = 92%

* Four levels of alloy with values: AISI 441, Alloy 800H, Kanthal A1 and P91.

(instead of the mean values used above) shows a similar consistency for n. The variability of the three sets of results can be considered as a measure of robustness. The variability in the three sets of results for log k_n is greater for P91 and Kanthal A1 than for AISI 441 and Alloy 800H as the plot in Fig. 12.31 shows. The differences between alloys are statistically significant. The variation between alloys accounts for the majority (61%) of the variation, which suggests that the laboratories are giving similar results to each other. There is no significant difference in robustness between long and short dwell experiments.

From this statistical analysis one can conclude that the proposed cyclic oxidation test procedure is robust. Any variation is dominated by alloy effects, thus the test procedure can separate different alloy behaviours. Only one alloy showed any dependency on the test laboratory. This was P91 and this laboratory interdependence may well reflect different heat treatment conditions for this alloy. Thus by testing to the guidelines, this round robin test has shown all laboratories can be expected to obtain similar results, within alloy reproducibility.

12.6 Prediction of alloy oxidation behaviour

The oxidation data can be modelled for each alloy and used to give an expected value with confidence interval when this Code of Practice is followed. The residuals from the model are reasonably random and approximately normally distributed implying that the model is satisfactory.

For log(k_n), using just the 19 values for AISI 441, Alloy 800H, Kanthal A1 and P91, the expected values and confidence intervals are given in Table 12.9.

12.7 Conclusions

The similarity of the results obtained by the various laboratories in this round robin implies that doing experiments according to this Code of Practice leads to consistent and robust results for these alloys. The consistency is

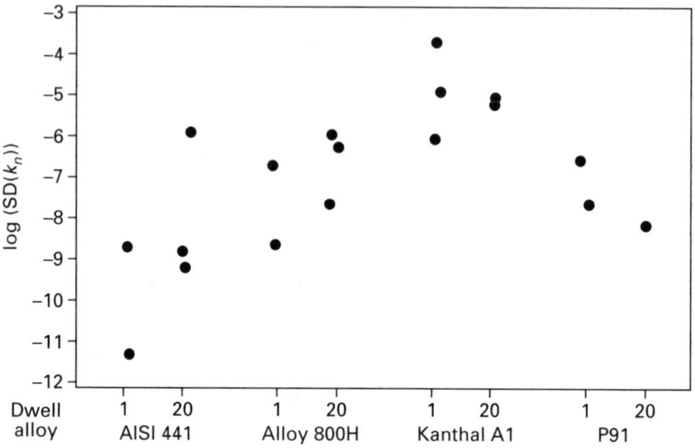

12.31 Scatterplot of log Sd(k_n) vs alloy.

Table 12.9 Expected values and confidence intervals for the oxidation rate constant (tabulated as log(k)) for alloys AISI 441, Alloy 800H, Kanthal A1 and P91 at the reference temperatures studied

Alloy	Test temperature (°C)	Lower 95% confidence interval	Expected log(k)	Upper 95% confidence interval
AISI 441	850	1.86	2.36	2.87
Alloy 800H	1000	5.24	5.80	6.35
Kanthal A1	1250	5.78	6.33	6.89
P91	650	2.76	3.48	4.19

similar for both long and short dwell experiments and thus the procedure can equally be applied to any hot dwell time, although it is preferred that 1 h (short) and 20 h (long) are adopted. The code is therefore both robust and effective in giving reliable results for cyclic oxidation testing as illustrated using the results from AISI 441, Alloy 800H, Kanthal A1 and P91. The picture is not so clear for P91 but this may be due to a variation in performance with alloy heat treatment for this alloy. Statistical analysis for CM 247 according to the scheme applied to the other alloys was not possible as the net mass change curves for this alloy are rather different from the others and the determination of the observables n, k_n, $t_{protective}$ and $\Delta m_{net}(t_{300\,h} - t_{protective})$ were not possible. However, also for CM 247 the code of practice enables reliable and reproducible results under thermal cycling testing to be obtained, as can be seen in Figs 12.12 and 12.13.

13

Final remarks

M. SCHÜTZE and M. MALESSA,
DECHEMA e.V., Germany

13.1 Summary

In the joint European COTEST project a solid basis has been developed for a reliable and meaningful standard for high-temperature cyclic oxidation testing. This basis combines scientific approaches with industrial needs and for the first time allows an intercomparison of data from different laboratories.

The evaluation of the existing test procedures and experimental facilities, together with the evaluation of the existing experimental data, showed that although a broad data basis existed, only a few directly comparable data sets were available underlining the real need of a standardised test procedure for cyclic oxidation (or thermal cycling oxidation) testing. The evaluation highlighted the parameters that had to be assessed more closely in the subsequent development of a set of test procedures covering a wide range of industrial applications.

Three different test types (long dwell times, short dwell times and ultra-short dwell times) were selected in such a way that together they cover practically the entire range of service conditions in which high-temperature materials are subjected to thermal cycling. In order to limit the number of parameters and to facilitate the data comparison, details of preparation of the corrosion test specimens, experimental procedures and data analysis were fixed in agreement. This set of internally standardised test procedures represented the basis for the testing programme of the reference materials.

The reference materials were subjected to the test procedures with defined variation of the test parameters. The main outcomes were:

- a scientific definition of the phase of a thermocycle;
- a minimum testing duration (accumulation of hot dwell time) of 300h;
- definition of key descriptors for the net mass change of specimens exposed to cyclic oxidation;
- evaluation of the significance of the different test parameters' influence on the key descriptors by statistical methods.

It turned out that the cyclic oxidation behaviour described by these parameters is very much dependent on the type of material used, i.e. whether Fe-Cr-spinels, chromia or alumina scales are formed. This can be regarded as an indication that the test procedure is powerful enough to distinguish between oxide scales of different protective potential.

Based on the outcome of the experimental studies in combination with the evaluation of the existing data and test procedures, a draft Code of Practice for thermal cycling oxidation testing was developed. This Code of Practice was formulated along the guidelines used for creating ISO test standards.

All principal contractors and a number of subcontractors were involved in the experimental validation of the draft Code of Practice. The tests were organised in such a way that each test was carried out in several laboratories. After testing the contractors were requested to supply the outcome of the materials behaviour by

- an exact record of all test parameters during the course of the test;
- gravimetric data for specimens as a function of time;
- gravimetric data for spalled oxide as a function of time;
- metallographic examination of corrosion damage;
- characterisation of corrosion products by electron microscopy on selected specimens.

Since all tests had to follow the guidelines developed in the COTEST programme it was expected that no large scatter should occur between the results from the different laboratories. It turned out at the end of this work package that indeed in most cases there was very little scatter between results from different laboratories and that this scatter could be explained mainly by the different cooling rates inherent to the different furnaces. However, even the duration of the protective time range and the behaviour in the non-protective time range in many cases showed a reasonable degree of agreement between the different laboratories. It was therefore concluded that the guidelines developed so far are sufficiently powerful for industrial use in order to lead to reliable data with a good degree of intercomparison between different laboratories.

Parallel to the completion of validation testing, the fine-tuning of the Code of Practice was completed. This was mainly directed towards additional details in the test procedure which were detected as significant in validation testing.

13.2 Conclusion

Prior to the COTEST project the lack of a standardised test procedure for thermal cycling oxidation testing hindered the intercomparison of experimental results between different facilities as in most cases the test parameters differed between the respective facilities.

The Code of Practice, which represents the main outcome of the project, demonstrated its robustness in the validation testing phase. Any variation is dominated by alloy effects; thus the test procedure can separate different alloy behaviours. Only one alloy showed a certain dependency on the test laboratory. This was heat-treated P91 and this laboratory dependency may reflect different heat treatment conditions for this alloy that had to be heat treated in batches due to technical constraints. Thus by testing to the guidelines, this round robin test has shown that all laboratories can be expected to obtain similar results, within alloy reproducibility.

The final version of the Code of Practice has already been submitted to Work Group 13 of ISO Technical Committee 156. It is expected that the revision of the Code of Practice by ISO TC156/WG 13 which commenced editing in November 2005 was completed by the end of 2006. The ISO working document is presently being processed within the usual ISO document flow and it is expected that the Code of Practice will be published as an international standard by ISO in the near future.

Appendix

Final Code of Practice – test method for thermal cycling oxidation testing

M. SCHÜTZE (on behalf of Working Party 3),
DECHEMA e.V., Germany

1 Scope

This draft Code of Practice describes the methodology for thermal cycling oxidation testing (known as cyclic oxidation testing) of metallic materials in gaseous atmospheres between ambient and elevated temperatures (series of measurements on a single test piece with repeated, regular and controlled temperature cycles). It may be applicable to other oxidisable materials with some modifications.

2 Normative references

This document was developed under the 'COTEST' project, which was funded by the European Commission within the fifth framework programme under EC contract no. G6RD-CT-2001-00639. The Code of Practice for discontinuous corrosion testing in high temperature gaseous atmospheres developed within the 'TESTCORR' project which was funded by the European Commission's 'Standards, Measurement and Testing Programme' (CEC contract no. SMT4-CT95-2001) serves as a general basis for this draft Code of Practice for cyclic oxidation testing. The present draft Code of Practice for cyclic oxidation testing follows where applicable the TESTCORR recommendations as close as possible with certain amendments and modifications described in the following sections.

As well as this standard for 'thermal cycling oxidation testing', standards for the following 'high-temperature corrosion testing methods are presently being developed in the frame of the work of ISO TC156 WG 13:

- *Thermogravimetric testing: in-situ* mass measurements at elevated temperatures on a single specimen without intermediate cooling.
- *Continuous isothermal exposure testing:* single post-exposure mass measurement on a series of specimens without intermediate cooling.
- *Discontinuous isothermal exposure testing:* series of mass measurements

on a single specimen with intermediate cooling at predetermined times not necessarily regular.

Other relevant publications:

- ASTM B76 – 90 (Reapproved 2001), Standard Test Method for Accelerated Life of Nickel–Chromium and Nickel–Chromium–Iron Alloys for Electrical Heating
- ASTM B78 – 90 (Reapproved 2001), Standard Test Method for Accelerated Life of Iron–Chromium–Aluminum Alloys for Electrical Heating
- S.R.J. Saunders, J.R. Nicholls, *Materials at High Temperatures* 1995, **13**, 115–120, 'Hot-salt corrosion testing – an international comparison'
- VAMAS – Versailles agreement on Advanced Materials and Standards – 'Hot salt corrosion testing', see special issue of *High Temperature Technology* 1989, **7**(4) and summary; S.R.J. Saunders 'Corrosion in the presence of melts and solids', pp. 85–103, in EFC Publication 14 'Guidelines for Methods of Testing and Research in High Temperature Corrosion' (1995), [Eds. H. J. Grabke and D. B. Meadowcroft]
- G54-84(1996) (Withdrawn 2002), Standard Practice for Simple Static Oxidation Testing, an updated version is currently being developed under ISO
- B. Tomkings, J.R. Nicholls, D.G. Robertson, EC Report EUR 19479 EN (2001), 'Discontinuous Corrosion Testing in High Temperature Gaseous Atmospheres "TESTCORR"'

3 Definitions

The definition of the main terms used in this Standard shall be as follows

3.1 Scale

Surface film and corrosion products produced on the surface of test piece by high-temperature corrosion.

3.2 Adherent scale

Scale adhering to the test piece even after cooling.

3.3 Spalled scale

Scale flaked from the test piece.

3.4 Spall

Collected spalled scales.

3.5 Metal loss

Cross-sectional reduction of sound metal.

3.6 Delaminated scale

Scale fully or partially detached from the surface but still in contact with the test piece.

3.7 Gross mass change

Mass change of test piece after cooling including collected spall.

3.8 Net mass change

Mass change of test piece after cooling without including the mass of spall.

3.9 High temperature corrosion

Corrosion occurring when the temperature is higher than the dew point of aqueous phases of the environment but at least 373 K. Specific to ultra-short cycle testing using Joule heating.

3.10 Initial resistance

Resistance calculated during the first heat cycle (measured at this specified temperature) by dividing the measured voltage drop due to the wire resistance by the current flowing through the wire.

3.11 Minimum resistance

The minimum value of resistance measured in the early few cycles. This results from thermal annealing of the wire/foil materials resulting in a reduction in material resistivity, thus resulting in resistance values lower than the initial resistance.

3.12 Resistance change

The measured change in resistance between that measured after any cycle and the minimum resistance.

4 Test apparatus

4.1 Design of apparatus

The apparatus shall be composed as a whole of the temperature regulating device for heating the test piece allowing thermal cyclic operation. The heating device should ideally be equipped with a testing portion capable of separating the test piece from outside air (this assembly is referred to as a closed system), unless this is impracticable for the cyclic test planned. Where applicable, a humidifying regulator should be used to continuously supply the gas kept at a constant humidity which should be monitored with a hygrometer. The gas supply shall be controlled by a gas flow meter.

An example of a basic design of a closed, horizontal, movable apparatus is shown in Fig. A.1. Other designs may use vertical orientation or moveable test piece supporters.

4.1.1

The test piece chamber shall not be composed of a material that reacts with the test atmosphere during the test to a degree that it changes the composition of the atmosphere.

4.1.2

The heating device should be designed such that the test piece chamber be isolated from the external environment. It should also be ensured that a continuous gas flow within the prescribed range passes the test pieces.

4.1.3

If a closed system with a test piece chamber cannot be used, then the tests may be performed in an open system with laboratory air. In this case the humidity of the air shall be recorded and the laboratories should be kept free from temperature changes and influences from weather conditions as far as possible. Ideally, however, closed systems should be used.

4.1.4

The furnace shall be characterised prior to the testing to determine the temporal temperature profile at a position near to the test piece. This can be achieved by using dummy test pieces and appropriate thermometry.

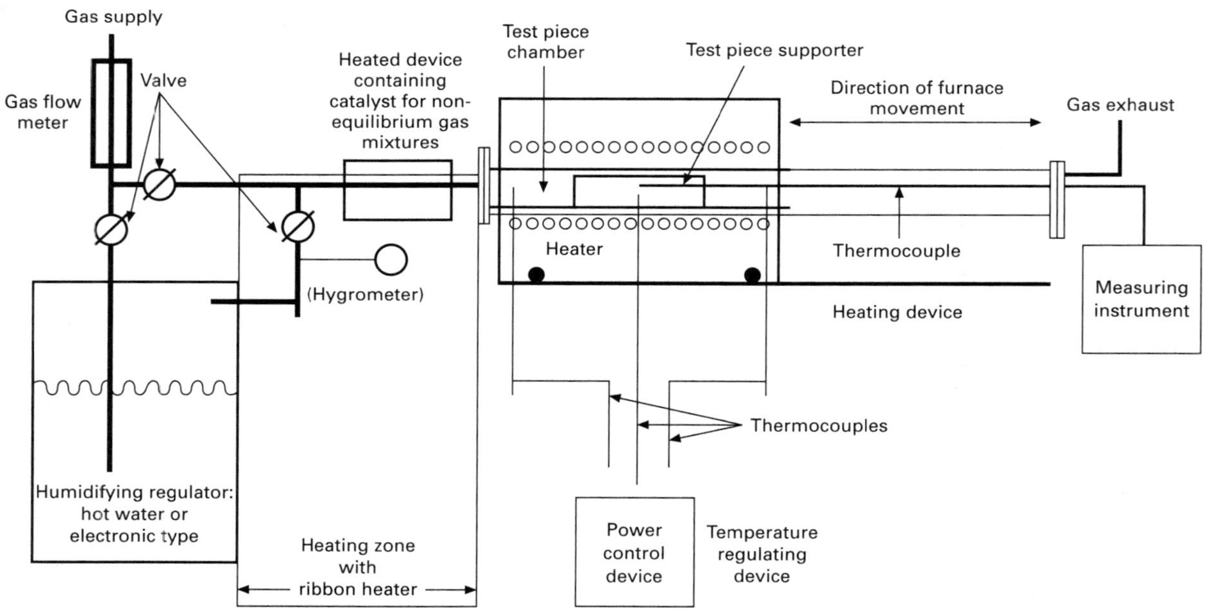

A.1 Basic design of a closed horizontal apparatus.

4.1.5

The temperature regulating device shall be capable of guaranteeing that the temperature of the test piece is kept within the permissible range given in Table A.1. Temperature variations/fluctuations due to movement of the furnace (less pronounced when test piece supports are moved) shall be compensated so that the temperature inside the furnace reaches the desired value rapidly without exceeding this temperature.

4.1.6

Heating device thermometry: when thermocouples are used, the material for the thermocouple shall withstand fully the test temperature and environment. Moreover, the diameter of wire is recommended to be as small as possible, within the limit where the thermoelectric power does not change in service. When pyrometers are used, the pyrometer must be focused within the cross-section of the test piece to be measured. A correction for the emissivity of the test piece must be applied.

4.2 Design of apparatus – ultra-short dwell tests – lamp furnace

Methods applicable to the long and short dwell tests may be used for ultra-short dwell testing, but a low thermal mass lamp furnace, or laser heating, must be used to heat the samples. The sample mass should be small to ensure rapid heating to the test temperature in a period as short as 1 min.

4.3 Design of apparatus – ultra-short dwell tests – Joule heating

Wire ribbon or foil samples may be heated by controlled electric power to permit rapid cyclic oxidation with durations of 1–10 min. One example of the Joule heating apparatus is illustrated in Fig. A.2 and consists of two terminal posts arranged to support a 'U'-shaped wire, ribbon or foil specimen. Joule heating results from the control of power to the wire, ribbon or foil sample. The temperature of the wire is measured using a 'calibrated' pyrometer.

Table A.1 Permissible tolerance of temperature of test piece

≤ 573 K	573 K < T ≤ 873 K	873 K < T ≤ 1073 K	1073 K < T ≤ 1273 K	> 1273 K
±2	±3	±4	±5	±7

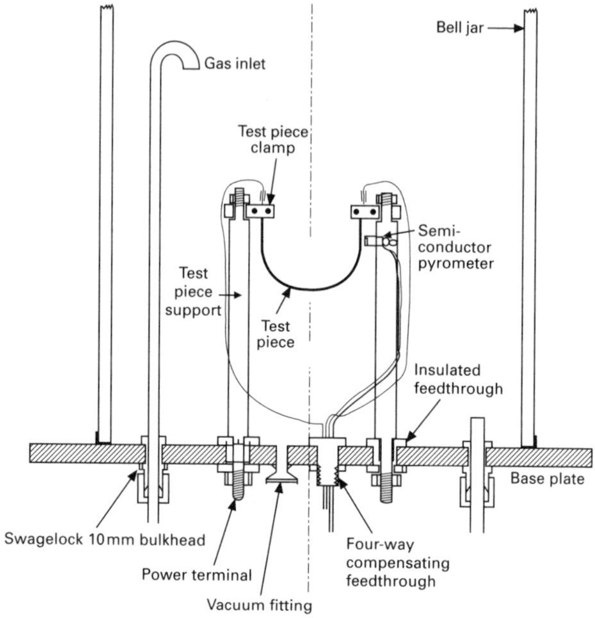

(a)

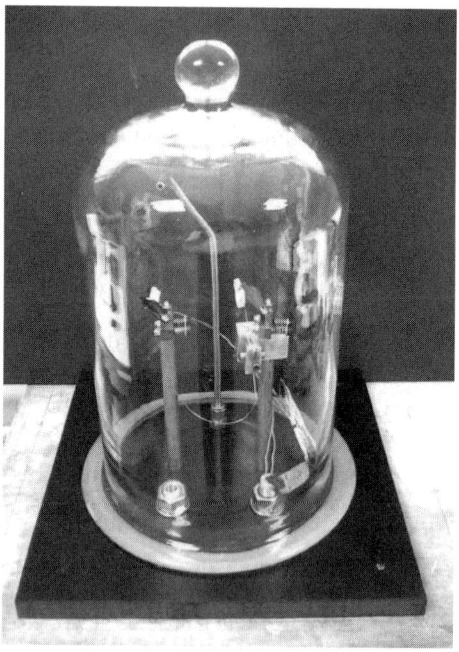

(b)

A.2 Basic design of a Joule-heated rapid thermal cycling apparatus: (a) schematic of apparatus; (b) apparatus with bell jar environmental chamber.

Measured temperatures are used to control the power to the sample, either manually or automatically. The apparatus is designed to allow control of environment using similar methods to those for long and short dwell cyclic oxidation tests.

4.3.1

Alternative design of apparatus, similar to those proposed in ASTM B76-90 and ASTM B-78-90, may be used for rapid cyclic oxidation testing in air. In this case environmental monitoring should follow the recommendations for long and short dwell tests in air.

4.3.2

The two terminals are attached to substantial binding posts that provide the current input for Joule heating.

4.3.3

Two additional conductors (wires) are attached to the free ends of the sample under test (see Fig. A.2a) to provide a four-point measurement system, from which voltage drop data can be measured. Together with control of the current supplied, this permits resistance to be measured and power to be controlled.

4.3.4 Power supply

The power supply should be capable of voltage, current or power control, being able to deliver voltages between 10 and 30 V to the circuit, with a continuous current capacity of at least 20 A/specimen. Figure A.3 illustrates the electrical circuit diagram for the Joule heating apparatus.

4.3.5 Voltage, current or power control

The preferred method is power control. Whichever mode of control is used the control circuit must be able to maintain control within ±0.5%.

4.3.6 Variable transformer

The transformer shall be capable of adjusting the voltage across the specimen so that current control is approx 0.25% of desired value and should have a continuous rating of 25 A.

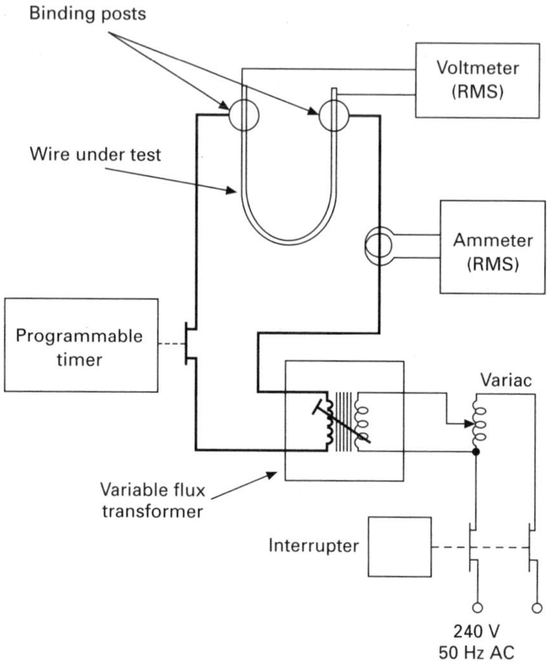

A.3 Electrical circuit diagram for Joule heating apparatus.

4.3.7 *Ammeter and voltmeter*

These instruments should have an accuracy of ±1%, when measuring nominal voltages and currents (typically 15 V and 15 A). If measuring AC supplies these should be true root mean square (RMS) meters.

4.3.8 *Optical pyrometer or infrared thermometer*

The optical system shall be such as to provide a magnification of at least ×10. The instrument must have an accuracy of ±5 K (±9 °F) and National Institute of Standards and Technology, Washington (NIST) traceability.

4.3.9 *Interrupter*

Some form of apparatus shall be used to interrupt the power supply, to open and close the circuit. Access to the specimen should only be possible with the power supply in the open condition (interrupted). On test completion, or wire failure, the power supply should revert to the open condition.

4.3.10 *Environmental containment*

The apparatus should include an environmental cabinet (a bell jar is the simplest design; see Fig. A.2b) to completely surround the specimen. This will limit operator access once the system is live (powered up), stop the influence of draughts for air operation and provide a facility for ensuring a controlled environment around the wire under test.

4.3.11 Environmental control

The apparatus should be fitted with the ability to remove the atmosphere by pumping to soft vacuum (10^{-2}mbar pressure) and back-filling with an inert gas or controlled environment. The controlled environment should be introduced to the top of the environmental chamber (bell jar) with gas flow arranged to occur by downward displacement, extracted at the base of the environmental chamber. Some indication of gas flow should be incorporated in the system.

4.3.12

An apparatus for recording, number of cycles to burn out (wire failure) or time to burnout should be incorporated.

4.4 Temperature monitoring

4.4.1

The temperature shall be measured by a suitable device according to ASTM E633-00. Thermocouples of type S (Pt – 10%Rh/Pt) or type R (Pt – 13%Rh/Pt) are preferred for temperatures ranging from room temperature up to 1700 °C. A thermocouple should be positioned close to the test piece surface and must be calibrated according to 4.4.2. If, however, the environment does not allow the use of such thermocouples in this way, the test piece temperature during one complete temperature cycle has to be deduced from the furnace calibration using dummy test pieces and appropriate thermometry in an inert environment.

4.4.2

Calibration of thermocouples shall be performed in accordance with ASTM E220-02 ('Standard method for calibration of thermocouples by comparison techniques'), ASTM E230-02 ('Standard temperature-electromotive forces tables for standardized thermocouples'), ASTM E1350-97 ('Standard test method for testing sheathed thermocouples prior to, during and after installation'). In this case, a representative thermocouple taken up from the batch of wire may be calibrated. It is recommended that they therefore be calibrated at the beginning and the end of each experiment if there is uncertainty about thermocouple stability.

4.4.3

The thermocouple shall be capable of confirming the temperature of the test piece to be within the range given in Table A.1. It has to be on a defined, fixed place as close to the test pieces as possible. In addition, it shall be adequately shielded in order to avoid radiation heat from the furnace wall.

4.4.4 Ultra-short dwell testing using Joule heating

During a Joule heating experiment, ultra-short dwell tests, it is very important that the temperature of the test specimen be adjusted as accurately as possible. For rapid cyclic oxidation it is required to run the test at a constant upper dwell temperature (±10 °K). This requires continual adjustment of the power to the wire throughout the test, either manually or automatically. One should note that the lifetime of a specimen depends critically on the test temperature (decreasing exponentially with reciprocal absolute temperature).

4.5 Gas supply for closed system operation

4.5.1

The gas supply system shall be capable of supplying the test gases at a constant rate to the test piece chamber (see example in Fig. A.1).

4.5.2

When a humidifying regulator is used it shall be capable of adjusting to the desired humidity. Deionised water of conductivity less than 1 μS/cm shall be used, unless otherwise specified.

4.5.3

The temperature of the space between humidifying regulator and test piece chamber shall be kept above the dew point in order to avoid condensation.

4.5.4

The gas flow shall be monitored by a gas flow meter. The flow meter shall be located as close as practicable to the inlet of the test piece chamber except where a humidifying regulator is used, in which case it shall be located upstream to the humidifier.

4.5.5

In the case that the gas is humidified the water vapour content shall be measured. This can be achieved by, for example, the use of a hygrometer upstream to the test piece chamber or measuring the condensed amount of water downstream to the chamber.

5 Test pieces

5.1 Test piece size and shape (long dwell and short dwell test)

5.1.1

The test piece shall, as a rule, be either a rectangular plate, a disc, a rod or a cylinder with a minimum surface area of 500 mm^2. Test piece geometry is an important factor and the following geometries are recommended:

- The rectangular plate test pieces should have dimensions of 17 mm length, 12 mm width and minimum 1.5 mm thickness or a surface area of at least 500 mm^2.
- The disc-shaped test pieces should have dimensions of 16 mm diameter and minimum 1.5 mm thickness or a surface area of least 500 mm^2.
- The rod-shaped test pieces should have dimensions of 50–120 mm length and 6–8 mm diameter or a surface area of least 500 mm^2.
- The cylindrical test pieces should have dimensions of 12 mm length and 10 mm diameter or a surface area of least 500 mm^2.

5.1.2

If the test pieces in 5.1.1 cannot be made, the shape and dimensions of the test piece shall be in accordance with the agreement between the parties concerned with the test results.*

5.2 Test piece size and shape (ultra-short dwell tests, using Joule heating)

5.2.1

Test pieces for cycling with ultra-short hot dwell times (see definition in 6.1) shall have a small thermal mass to allow rapid heating and cooling of specimens. Therefore wires or foils may be used. Recommended dimensions for wires

* When thin materials are tested, the presence of sharp edges shall be taken into account which may lead to significantly differently oxidation behaviour compared with thicker specimens.

are 0.4–0.7 mm in diameter and at least 150 mm in hot length. The preferred length is 220 mm. The dimensions of the foils shall be 30–100 μm in thickness, at least 150 mm in length and 1.25–5 mm in width. Recommended dimensions for foils are 30–100 μm thickness, 220 mm length and 3 mm width.

5.3 Test piece manufacture

5.3.1

The test pieces shall be finished by machining so that the layers affected by cutting do not remain.

5.3.2

The final finishing of the surface of the test pieces should be 1200 grit. Edges shall be rounded. No sharp edges must be present. If parties agree, different finishes may applied.

5.3.3

Other than 5.3.2, the surface finish condition shall be described.

5.3.4

The surface of the test pieces shall not be deformed by marking, stamping or notching. Identification of the test pieces shall solely be on the basis of recording the relative position within the test chamber.

5.3.5

If holes for test piece support are needed the distance from the edges shall be greater than the thickness of the test piece. The edges of the holes shall be dressed.

5.3.6

Details about specimens for burner rig and deposit testing are given in the respective annexes.

5.4 Materials/specimen characterisation before testing

5.4.1

The chemical composition of the materials shall be reported in terms of chemical analyses preferential to nominal composition. It is also recommended to store a piece of the particular tested reference material.

5.4.2

The method of heat treatment of material/test pieces shall be reported.[†]

5.4.3

It is recommended that the microstructure of the materials in a pre-test condition after preparation be determined according to ASTM E3-01 and etching according to ASTM E407-00.

5.4.4

The test pieces shall be degreased by ultrasonic cleaning in a suitable fresh solvent[‡]. Iso-propanol or ethanol is recommended. Alternatively the LEAFA[§] cleaning procedure may be applied for degreasing of test pieces. Once cleaned any contamination of the test piece has to be avoided.

5.4.5

Prior to weighing and exposure, the cleaned test pieces shall be stored in a desiccator (no longer than 72 hours).

5.4.6

The dimensions of the test piece shall be measured with a precision of ±0.02 mm by means of the measuring instruments specified in ISO 3611 and ISO 6906.

5.4.7

The mass of the test pieces shall be determined prior to exposure. Three measurements shall be made for each test piece. The standard deviation of

---

[†] Heat treatment of certain materials with a composition of the protective scale forming element at the lower limit ('borderline materials') may significantly influence the oxidation behaviour of these materials as diffusion pathways and mechanism of scale formation may be altered. This depends on the applied heat treatment procedure and has to be reported.

[‡] Chlorinated solvents shall not be used.

[§] LEAFA: Life Extension of Alumina Forming Alloys, EC Contract No. BRPR-CT97-0572; The Cleaning Procedure, published in: J. R. Nicholls, 'Discontinuous Measurement of High Temperature Corrosion', in Vol. 14 of the EFC Publication Series 'Guidelines for Methods of Testing and Research in High Temperature Corrosion', eds. H. J. Grabke and D. B. Meadowcroft, The Institute of Materials, London 1995. Guidelines for Methods of Testing and Research in High Temperature Corrosion' (1995) [Eds H J. Grabke and D.B. Meadowcroft]. The Institute of Materials, London.

the measurements shall not exceed 0.05 mg for test pieces with geometries described in 5.1.1.

5.4.8

The number of replicates used in the test shall be three or more.

6 Test method

6.1 Definition of a thermal cycle

6.1.1

A thermal cycle consists of the heating phase, the hot dwell time, the cooling time and the cold dwell time. An example for a hot dwell temperature of T_{dwell} = 1273 K is given in Fig. A.4. The four phases of a temperature cycle are defined in 6.1.2 to 6.1.5.

6.1.2

The heating time starts when the test pieces are heated, e.g. by entering the furnace, and ends with the beginning of the hot dwell time which is defined in 6.1.3.

6.1.3

The hot dwell time starts when the actual temperature exceeds 97% of the desired hot dwell temperature T_{dwell} (measured in K). Extensive numerical

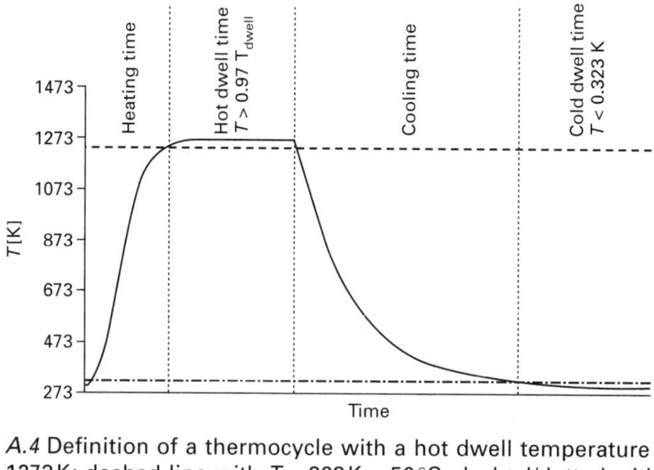

A.4 Definition of a thermocycle with a hot dwell temperature of 1273 K; dashed line with T = 323 K = 50 °C; dashed/dotted with $T = 0.97\ T_{dwell}$ = 1234 K.

calculations and comparison between hypothetical and real temperature cycles have shown that only those times of the temperature cycle contribute to oxidation of the test pieces where the temperature is close to the hot dwell temperature (see Chapter 5). The hot dwell time ends upon removal from the furnace.

6.1.4

The cooling time starts when the heating of the test piece is stopped, e.g. by the removal of the test piece from the furnace, and ends when the actual test piece temperature falls below 50 °C.

6.1.5

The cold dwell time starts after the test pieces have cooled below 50 °C and ends when the test pieces are heated again.

6.2 Types and dwell times of thermal cycles

6.2.1

Three general types of thermal cycles are typical for industrial applications. *Long dwell time testing* aims to simulate conditions in large-scale industrial facilities encountered in applications such as power generation plants, waste incineration or chemical industry. In these applications the metallic components are designed for extremely long-term operation, e.g. for typically up to 100 000 h. Thermal cycling of materials occurs to planned plant shut-downs, e.g. for regular maintenance, or to unplanned shut-downs as a result of offset conditions. Therefore, the time intervals between various thermal cycles are relatively long and the number of cycles is, relative to the long operation time of the components, comparatively small, i.e. typically around 50 cycles.

Thermal cycling with *short dwell times* is typically experienced in applications such as industrial gas turbines, jet engines, automotive parts, heat treatment facilities, etc. The intervals between start and shut-down of the facilities are generally much shorter than in applications with long dwell times. Also, the design life and/or the time until complete overhaul/repair (typically 3000–30 000 h) are shorter and, depending on the specific practical application, the number of cycles is much higher than in the cases above.

Testing with *ultra-short dwell times* mainly addresses applications of high-temperature alloys as heating elements in the form of wires or foils. Another typical application in which such short cycles prevail would be catalyst foil carriers, e.g. in cars. In such applications the number of cycles is, related to the overall design life (typically several hundred to a few thousand hours), extremely high and the time intervals between heating and cooling can be as low as minutes or even seconds. Such conditions are commonly encountered in a number of industrial applications such as, for example, burners and hot gas filters, but also in a large variety of domestic applications where metallic heating elements are used, e.g. in cooking plates, toasters, boilers, dryers, fryers.

6.2.2 *Dwell time test parameters for long dwell time testing*

The definition of 6.1 for a thermal cycle shall be applied to long dwell time testing. For practical reasons it is useful to stay within a 24 h operation rhythm. Therefore experiments shall be performed with a 20 h hot dwell time and 4 h period which includes the cooling time, the cold dwell time and the heating time. The heating and cooling times according to 6.1.2 and 6.1.4 shall be reported.

6.2.3 *Dwell time test parameters for short dwell time testing*

The definition of 6.1 for a thermal cycle shall be applied to short dwell time testing. The hot dwell time shall be 1 h, the cold dwell time shall be 15 min. Heating and cooling times according to 6.1.2 and 6.1.4 shall be reported. For weighing procedures for mass determinations the cold dwell time may be extended but shall be kept as short as possible.

6.2.4 *Dwell time test parameters for ultra-short dwell testing*

The definition of 6.1 for a thermal cycle shall be applied to ultra-short dwell time testing. The hot dwell shall be 5 min and the cold dwell 2 min. It is not possible to report heating and cooling times according to 6.1.2 and 6.1.4 as the optical pyrometer cannot respond fast enough to changes in temperature during the heat and cooling cycle, when manually operated. The hot dwell time can be estimated from the change in resistance with temperature. The resistance change can be measured during the rapid heating cycle.

6.2.5 *Dwell time test parameters for testing under corrosive conditions*

Depending on the type of thermal cycle, the dwell time test parameters defined in 6.2.2 to 6.2.4 shall be applied. If this is not possible due to practical reasons, the testing conditions, in 6.6 shall be applied.

6.3 Testing duration

The testing duration shall be at least 300 h of accumulated hot dwell time to allow a significant oxidation of the test pieces. For more reliable results it is, however, recommended to extend the accumulated hot dwell time to at least 1000 h.

Note that for ultra-short dwell testing the lifetime may be controlled by the number of cycles. Lifetimes of less than 300 h are often observed.

Figure A.5 shows three types of different oxidation behaviour giving an indication of minimum testing duration. Testing is terminated when protective behaviour is no longer found for the materials.

6.4 Supporting of test pieces

6.4.1

The test pieces shall be supported according to the following principles.

6.4.2

The test piece shall be supported by a material that does not react at the test temperature. Contacts between test piece and support shall be minimised.

6.4.3

The supporter of the test piece to be used shall ideally be designed to be able to collect the scale if it flakes during testing or during cooling after finishing the test.

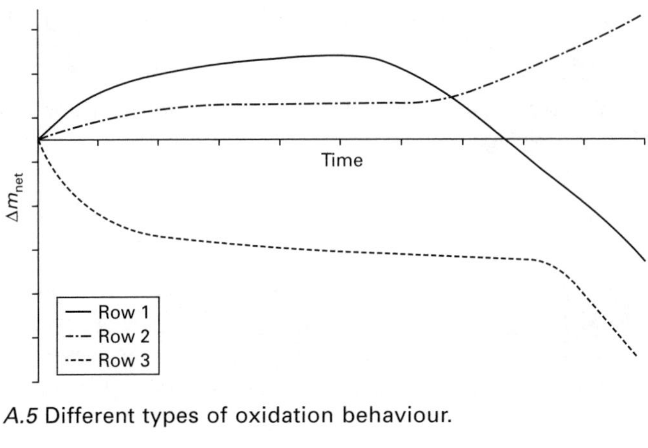

A.5 Different types of oxidation behaviour.

6.4.4

When spalled oxide is collected and multiple test pieces are exposed simultaneously, each test piece shall be inserted into an individual test piece supporter in order to enable the collection of all spall individually from each test piece.

6.4.5

The support design shall ensure that the test atmosphere flows evenly over the major faces of each test piece. This can be achieved by holes of 1–2 mm in length in width in the side wall of the test piece supporters in the bottom area.

6.4.6

Examples of suitable test piece supports and basic layout of test piece arrangement are shown in Figs A.6 to A.8.

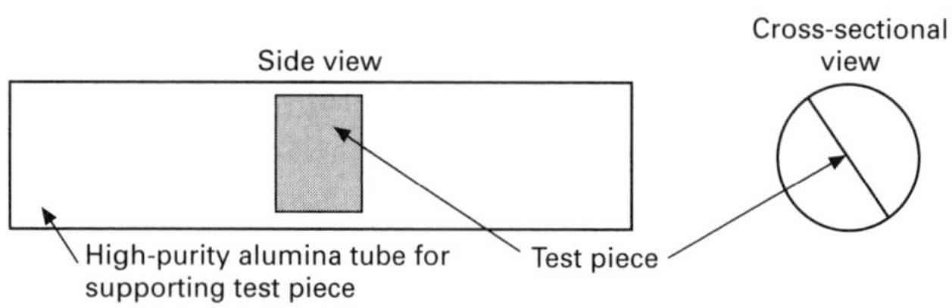

A.6 Test piece supporter and basic layout of test piece arrangement – tube design.

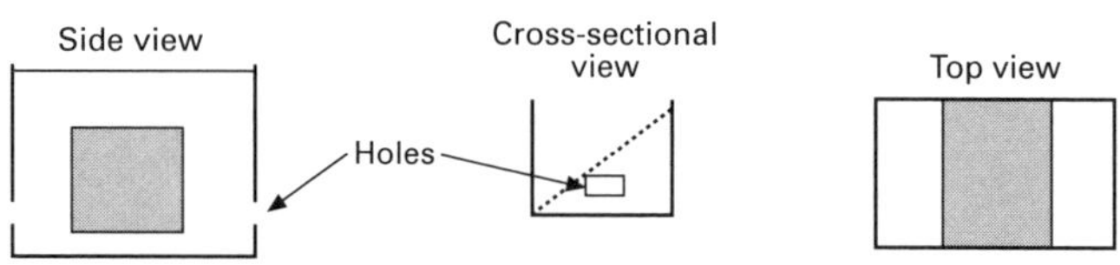

A.7 Test piece supporter and basic layout of test piece arrangement – U-shaped design.

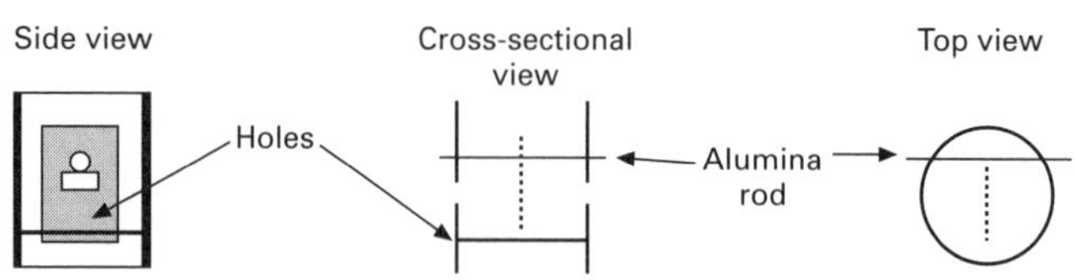

A.8 Test piece supporter and basic layout of test piece arrangement – rod supported design.

6.5 Test environment in air oxidation testing

6.5.1

The gas flow shall be high enough to ensure an overpressure at the exit of the furnace and that no depletion of reaction species will occur. At the same time the gas flow shall be slow enough to allow the gas mixture to preheat and (if applicable) in some applications to reach equilibrium. For tubes with a diameter between 5 and 15 cm, the volume flow rate shall be between 5 and 10 l of gas per hour, resulting in a gas flow velocity of 0.02–0.35 mm/s.

6.5.2

For COTEST dry synthetic air shall be used as test gas as humidity of laboratory air varies significantly depending on the location of the laboratory and local weather conditions. If tests have to be performed in humidified air, the water content has to be reported in specific humidity (g of water per kg air).

6.5.3

For COTEST dry synthetic air shall be used within the validation testing for all materials except for P91 which shall be tested in humidified air with a specific humidity of 20 g water/kg air corresponding to a dew point of 25 °C.

6.5.4

As synthetic air the following gas composition shall be used: 20 vol.% O_2, 80% vol.% N_2.

6.6 Special test parameters in complex corrosive environments

6.6.1 Test environment

Test environments for testing in complex corrosive atmospheres, under deposits or burner rig conditions, are given in the annexes to this Code of Practice.

6.6.2 Dwell time test parameters

Dwell time test parameters for testing in complex corrosive atmospheres, under deposits or burner rig conditions, are given in the annexes to this Code of Practice.

6.7 Determination of mass change by oxidation

The use of tweezers is recommended. Test pieces shall never be touched with the hands in order to eliminate any contamination (grease, salts). Care has to be taken when using gloves as the contamination with the separating agent of the gloves leads to falsification in mass determination. If consistency between individual measurements cannot be found, the measurement environment has to be controlled.

6.7.1 Measurements prior to testing

6.7.1.1

New test piece supporters shall be put into the furnace for at least 24 h to remove water and other residues from production. It is recommended that each empty test piece supporter to be used is subjected to one thermal cycle before the test to remove water traces.

6.7.1.2

The mass of the test pieces shall be determined prior to exposure. Three individual measurements shall be made for each test piece with a precision of 0.02 mg. The standard deviation for the measurements shall not exceed 0.05 mg.

6.7.1.3

The mass of the test piece supporters shall be determined prior to exposure. Three individual measurements shall be made for each test piece supporter with a precision of 0.02 mg. The standard deviation for the measurements shall not exceed 0.05 mg.

6.7.1.4

The mass of the test piece supporter containing one test piece shall be determined prior to exposure. Three individual measurements shall be made for each test piece supporter containing one test piece with a precision of 0.02 mg. The standard deviation for the measurements shall not exceed 0.05 mg.

6.7.2 Intermediate and final mass change determination

6.7.2.1

After removal from the furnace, the test piece supporters containing the test pieces shall be settled in the weighing room for 15 min to allow them to acclimatise. The test pieces shall not be descaled unless specified by the customer.

6.7.2.2

For each mass change determination at intermediate stages and final stage, the mass of the test piece supporter containing one test piece and spalled scales (the test piece supporter including spalled scales) and the mass of the test piece including adherent scales shall be determined as shown in Fig. A.9. Three individual measurements shall be made for each test piece supporter containing one test piece, each test piece supporter and each test piece with a precision of 0.02 mg for each measurement. The standard deviation for each set of measurements shall not exceed 0.05 mg.

6.7.2.3

Care must be taken during mass measurements to avoid spallation of oxide scales caused by mechanical contacts with tweezers, etc. A test piece supporter with a rod support design as shown in Fig. A.8 circumvents the problem and allows the determination of mass changes without direct contact of the tweezers with the test piece as shown in Fig. A.10.

6.7.2.4

Gross mass change Δm_{gross} as defined in 3.7 is determined according to Eq. 6.1:

$$\Delta m_{gross}(t_n) = m_{ST}(t_n) - m_{ST}(t_0) \quad 6.1$$

A.9 Mass determination – variant I.

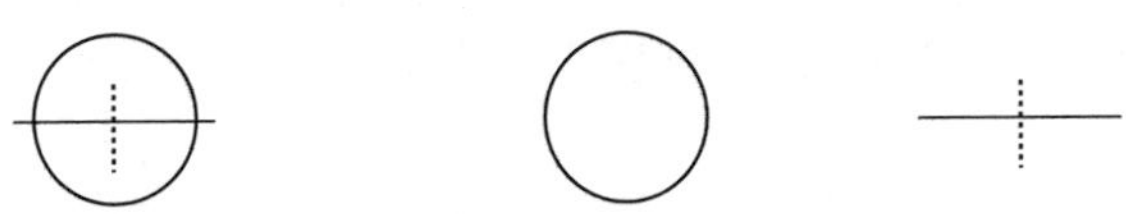

A.10 Mass determination – variant II (test piece can be removed by holding the rod with tweezers).

where $\Delta m_{gross}(t_n)$ = gross mass change at time t_n (mg),
$m_{ST}(t_n)$ = mass of test piece supporter and test piece at time t_n (mg),
$m_{ST}(t_0)$ = mass of test piece supporter and test piece prior to the test (mg).

6.7.2.5

The mass of spalled oxide Δm_{spall} is determined according to Eq. 6.2:

$$\Delta m_{spall}(t_n) = m_S(t_n) - m_S(t_0) \qquad 6.2$$

where $\Delta m_{spall}(t_n)$ = mass change of spalled material at time t_n (mg),
$m_S(t_n)$ = mass of test piece supporter at time t_n (mg),
$m_S(t_0)$ = mass of test piece supporter prior to the test (mg).

6.7.2.6

Net mass change Δm_{net} as defined in 3.8 is determined according to Eq. 6.3:

$$\Delta m_{net}(t_n) = m_T(t_n) - m_T(t_0) \qquad 6.3$$

where $\Delta m_{net}(t_n)$ = net mass change at time t_n (mg),
$m_T(t_n)$ = mass of test piece at time t_n (mg),
$m_T(t_0)$ = mass of test piece prior to the test (mg),

or when a rod-supported design as shown in Fig. A.8 is used by Eq. 6.4:

$$\Delta m_{net}(t_n) = \Delta m_{gross}(t_n) - \Delta m_{spall}(t_1) \qquad 6.4$$

where $\Delta m_{net}(t_n)$ = net mass change at time t_n (mg),
$\Delta m_{gross}(t_n)$ = gross mass change at time t_n (mg),
$\Delta m_{spall}(t_n)$ = mass of spalled material at time t_n (mg).

6.7.3 *Frequency of mass change determination*

6.7.3.1

For long dwell time testing mass change measurements shall be made once each working day for the initial part of the test, measurements may be made less frequently as the test progresses.

6.7.3.2

In short dwell time testing the extension of the cold dwell phase due to the mass change measurements must be taken into account. Therefore the number of mass change measurements shall be small enough to minimise these possible influences but high enough to obtain meaningful mass change curves. Mass change measurements on a daily basis have proven to be practicable.

6.7.3.3

In ultra-short dwell testing, pieces cannot be remounted in the test facility after mass change measurement. Therefore individual test pieces shall be removed after a range of cycles/exposure times to determine mass change and correlate this with the change in resistance while the rest of the initial number of test pieces continue to be subjected to thermal cyclic oxidation testing. Test pieces shall be removed after 10, 20, 40 and 100 h of accumulated hot dwell time.

6.7.3.4

It is recommended to document the macroscopic appearance of the test piece surface by macro-photographs.

6.7.4 Analysis of mass change

6.7.4.1

Net mass change of test pieces shall be plotted versus time as shown in Fig. A.11. According to the common use in high-temperature oxidation testing, the mass change is usually described mathematically by Eq. 6.5 (see also Fig. A.11). The values for the oxidation rate k, the exponent of the growth law n, the protective oxide growth time $t_{protective}$ and the corresponding number of cycles $N_{protective}$ shall be reported. Determination of these values is described in 6.7.4.2 to 6.7.4.4.

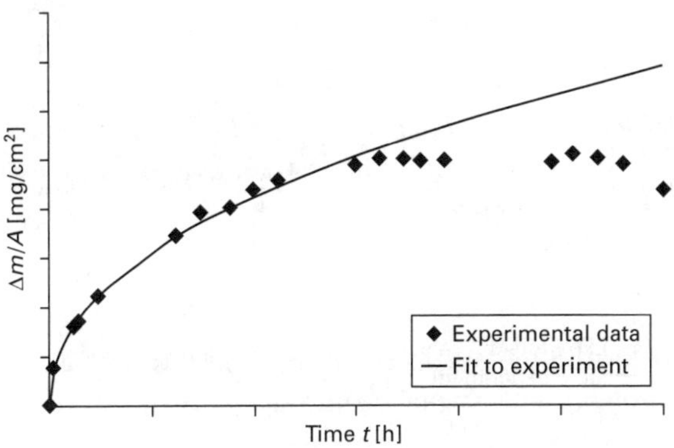

A.11 Net mass change as a function of time.

6.7.4.2

The double logarithmic plot (log net mass change vs log time) as shown in Fig. A.12 reveals a change in the mechanism of oxidation behaviour. As long as protective oxide growth takes place it can be described mathematically by Eq. 6.5.

$$\left(\frac{\Delta m_{\text{net}}}{A}\right)^n = kt \rightarrow \frac{\Delta m_{\text{net}}}{A} = k^{\frac{1}{n}} t^{\frac{1}{n}} \rightarrow \log\left(\frac{\Delta m_{\text{net}}}{A}\right) = \log\left(k^{\frac{1}{n}} t^{\frac{1}{n}}\right)$$

$$= \log k^{\frac{1}{n}} + \frac{1}{n} \log t \qquad 6.5$$

A change in the mechanism (spallation, breakaway oxidation) becomes apparent by a change of the slope in the double logarithmic plot. Analysis of the linear part of the curve by linear regression with means of simple spreadsheet calculations yields the slope $b = 1/n$ and the y-axis intercept as shown in Fig. A.12.

6.7.4.3

From the y-axis intercept the oxidation rate constant k for the linear region shall be calculated as given in Eq. 6.6:

$$a = \log k^{\frac{1}{n}} \rightarrow a = \frac{1}{n} \log t \rightarrow k = 10^{a \cdot n} \qquad 6.6$$

6.7.4.4

For the determination of k, n and the protective oxide growth time $t_{\text{protective}}$ an iterative procedure shall be applied. In a first step a best fit line shall be

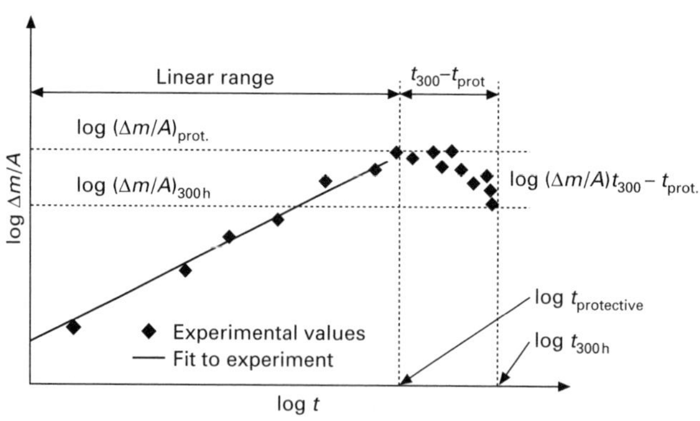

A.12 Double logarithmic plot of mass change vs time.

drawn through the data points in the 'linear range' using the functionality of simple spreadsheet calculations. The correlation coefficient defined by Eq. 6.7 shall be maximised by including or excluding data points near to the point where data points leave the linearity area:

$$\Delta m_{net}(t_{300\,h} - t_{protective}) = \Delta m_{net}(t_{300\,h}) - \Delta m_{net}(t_{protective}) \qquad 6.7$$

where $\Delta m_{net}(t_{300\,h})$ = net mass change of test piece after 300 h (mg), $\Delta m_{net}(t_{protective})$ = net mass change of test piece at $t_{protective}$ (mg).

with the standard deviations and and the covariance between two data sets defined by Eq. 6.8:

$$\Delta m_{net}(t_{1000\,h} - t_{protective}) = \Delta m_{net}(t_{1000\,h}) - \Delta m_{net}(t_{protective}) \qquad 6.8$$

where $\Delta m_{net}(t_{1000\,h})$ = net mass change of test piece after 1000 h (mg), $\Delta m_{net}(t_{protective})$ = net mass change of test piece at $t_{protective}$ (mg).

The data points included in the calculation of the correlation coefficients shall be used to calculate k and n. The latest data point included in the linear regression shall be reported as $t_{protective}$ (protective oxide growth time).

6.7.4.5

The number of cycles $N_{protective}$ that corresponds to the protective oxide growth time $t_{protective}$ and which is the last data point that is included in the linear regression shall be reported.

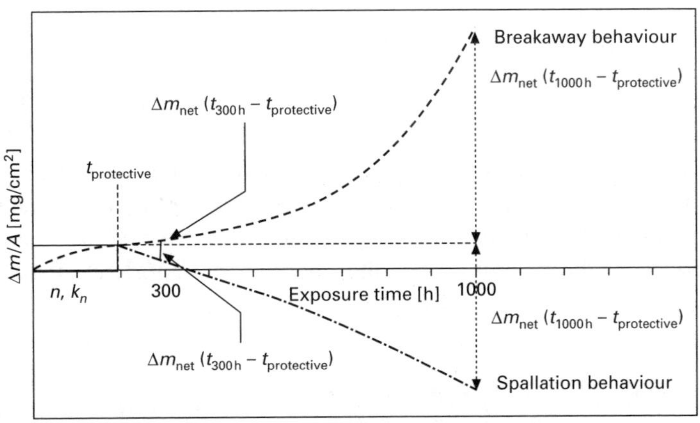

A.13 Net mass changes differences $\Delta m_{net}(t_{300\,h} - t_{protective})$ and $\Delta m_{net}(t_{1000\,h} - t_{protective})$ for materials showing spallation behaviour (dashed/dotted line) and breakaway behaviour (dashed line)

6.7.4.6

Additionally the following net mass changes difference shall be reported as given in Eq. 6.7 and 6.8 and shown in Fig. A.13.

6.8 Determination of resistance change due to oxidation (ultra-dwell tests)

6.8.1

During the hot dwell period of the first thermal cycle the current and voltage readings are adjusted to reach the required test temperature:

- The initial power (P_I) is calculated from $V_{rms} \times I_{rms,}$ assuming AC Joule heating.
- The initial resistance (R_I) is calculated from V_{rms}/I_{rms}, assuming AC Joule heating.

6.8.2

Control of the rapid cyclic oxidation test is through control of the power. The power level is adjusted to keep the upper dwell temperature constant, to within specified tolerance.

6.8.3 Frequency of resistance measurement

6.8.3.1

Resistance measurements should be determined every cycle until stabilised (stress relief during early cycles can lead to a drop in resistance). The stabilised, minimum resistance should be recorded (R_o).

6.8.3.2

During the ultra-short dwell cyclic oxidation tests, hot resistance should be determined periodically as necessary to keep power control of temperature within ±10 K. It is recommended that such measurements are taken every 30–50 cycles throughout the test until burnout, and more frequently if rapid changes in resistance are observed.

6.8.4 Analysis of resistance change

6.8.4.1

Fractional change in resistance $\left(\frac{\Delta R}{R_o}\right)$ is defined as

$$\frac{R(t_n) - R_o}{R_o} \qquad 6.9$$

where $R(t_n)$ is the measured hot resistance (V_{rms}/I_{rms}) at time t_n, measured in ohms, where n is the cycle number. R_o is the minimum resistance determined in or shortly after cycle 1 (see Fig. A.14 for definition). R_o may be less than R_I due to thermal annealing during the first few cycles.

6.8.4.2

Fractional change in resistance shall be plotted versus number of cycles (time) as shown in Fig. A.15. The fractional resistance change plot is a dimensionless measure of the effect of oxidation rate on the resistance of the wire and thus can normalise the performance of wires for different diameters. This is not the case for foil materials, where change in cross-sectional area must be taken into account.

6.8.4.3

The resistance of a wire, foil or ribbon sample is given by:

$$R = \rho\frac{L}{A} \qquad 6.10$$

where R = resistance, ρ = resistivity, L = length of the wire, foil or ribbon and A is the cross-sectional area.

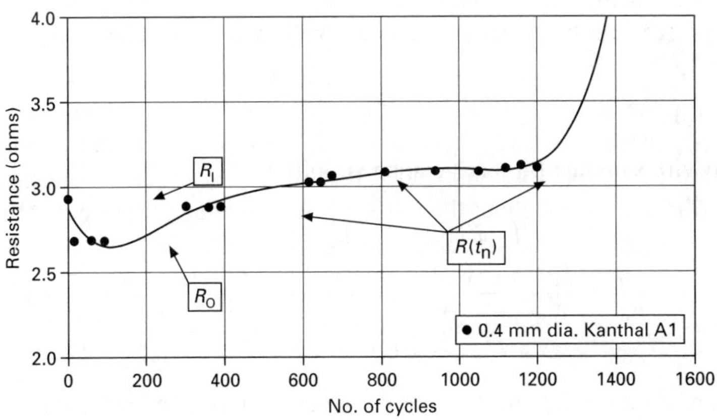

A.14 Definition of R_I, R_o and $R(t_n)$.

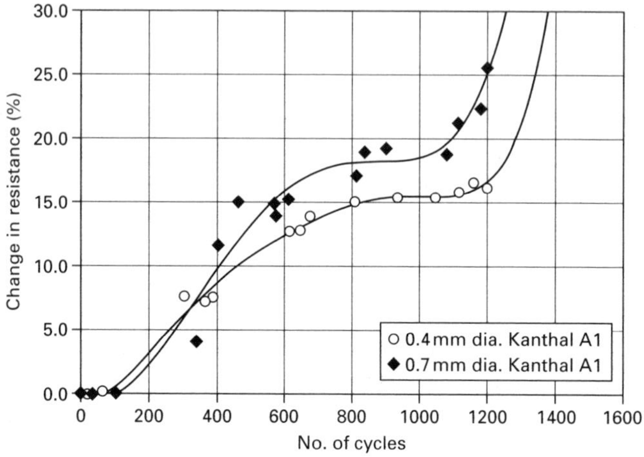

A.15 Change in resistance with number of cycles.

The fractional resistance defined in Eq. 6.9 can be related to specimen dimensions by:

$$\frac{\Delta R}{R_o} = \frac{R(t_n) - R_o}{R_o} = \frac{\rho \cdot \dfrac{L(t_n)}{A(t_n)} - \rho \dfrac{L_o}{A_o}}{\rho \dfrac{L_o}{A_o}} \qquad 6.11$$

where $L(t_n)$, $A(t_n)$ relate to the specimen length and cross-sectional area at time t_n (after n cycles), ρ is the wire resistivity for a thermally annealed wire, and L_o and A_o are the initial wire length and cross-sectional area. (Note that ρ may be less than ρ_o, the initial resistivity of the wire, due to thermal annealing effects over the first few thermal cycles).

Thus changes in wire cross-sectional area $A(t_n)$ at time t_n, due to oxidation processes, can be directly correlated with measured values.

6.8.4.4

For wire samples Eq. 6.11 can be simplified:

$$\frac{\Delta R}{R_o} = \frac{4\rho}{\pi} \cdot \frac{\left(\dfrac{L(t_n)}{(d(t_n))^2} - \dfrac{L_o}{(d_o)^2}\right)}{\dfrac{4\rho}{\pi}\dfrac{L_o}{(d_o)^2}} = \frac{L(t_n)}{L_o} \cdot \frac{d_o^2}{(d(t_n))^2} - 1 \qquad 6.12$$

where d_o and $d(t_n)$ are the wire diameters initially and at time t_n.

6.8.4.5

For small changes in wire dimensions this permits an estimate of metal loss from oxidised wires. The metal loss (Δx) would be $[d_o - d(t_n)]/2$. Thus equation 6.12 simplifies to

$$\Delta x \simeq \frac{d_o}{4} \cdot \frac{\Delta R}{R_o} \qquad 6.13$$

Thus a plot of versus time can be treated in a similar manner to mass change data (6.7.4) to study oxidation kinetics for the early stages of oxidation where the assumptions of small section loss hold true. For a more exact calculation or for foil and ribbon samples Eq. 6.11 has to be used to calculate changes in cross-sectional area.

7 Post-test evaluation of test pieces

7.1 Macroscopic evaluation

The macroscopic appearance of the test piece surface shall be photographed.

7.2 Metallographic cross-section

7.2.1

Scale thickness measurements shall be made by metallographic cross-sections following the TESTCORR recommendations. Care has to be taken in mounting the specimen orthogonal to the primary axis of the test piece. The cross-sectioned test pieces shall be analysed using conventional light microscopy. Measurements shall consist of:

- deposit thickness;
- scale thickness;
- depth of internal penetration;
- depth of grain boundary attack measured;
- depth of any depleted zone;
- remaining cross-section of unaffected material.

A minimum of eight measurements per test piece shall be obtained. In addition, the position of maximum attack shall be measured.

8 Report

8.1 Matters to be described

The following data, where known, shall be included in the report on the test results.

8.1.1 Test material

- Manufacturer
- Name of material (manufacturer designation; ASTM, DIN, etc.)
- Grade or symbol
- Heat number/batch number
- Chemical composition
- Processing condition
- Heat treatment condition

8.1.2 Test piece

- Designation of test piece
- Dimensions and surface area of test piece
- Surface finish condition of test piece
- Degreasing method of test piece
- Method of test piece support
- initial mass

8.1.3 Testing environments

- Test temperature and maximum and minimum temperatures during the test
- Test duration
- Hot dwell time, cold dwell time, heating time, cooling time
- Dew point temperature of humidified air or humidity of laboratory air
- Flow rate of test gas
- Volume flow rate of test gas
- Open or closed system according to 4.1.3

8.1.4 Test result

- Plot of mass change per area in mg/cm^2 vs time according to 6.7.2
- Results of any metallographic investigations performed according to 7.2.1
- Amount of spalled scale in mg
- Photograph of appearance after testing
- Photograph of cross-section including the surface layer of the metallographic section of test piece after testing
- Oxidation rate k, exponent of the growth law n, cycles N_z and the accumulated hot dwell time t_z that corresponds to the onset of spallation or breakaway oxidation according to 6.7.2

8.2 Supplementary note

It is desirable additionally to describe the following matters in the report on the test results:

8.2.1

Mechanical properties of the raw material at room temperature

8.2.2

Sampling conditions of the test piece from raw material

8.2.3

Outline of the test apparatus

8.2.4

SEM/EDX analysis

9 Annex A: Thermal cycling oxidation testing with deposits

A.1 Scope

A.2 Application of Code of Practice

Sections 1 to 5 of the Code of Practice shall apply.

A.3 Test method

Section 6 of the Code of Practice shall apply with the following modifications:

A.3.1 Dwell time of the thermal cycle

The methodology of *long dwell time testing* shall be applied (Section 6.2 of the Code of Practice).

A.3.2 Dwell time test parameters

The definition of 6.1 of the Code of Practice shall apply. For practical reasons, the 24 h operation rhythm is maintained with a hot dwell time of 18 h and a 6 h period, including the cooling time, the cold dwell time and the re-heating time. The heating and cooling times according to 6.1.2 and 6.1.4 of the Code of Practice shall be reported (Section 6.2.2 of the Code of Practice).

A.3.3 *Testing duration*

For COTEST, the minimum number of cycles for deposit testing shall be 17, giving a hot dwell duration total of 306 h (Section 6.3 of the Code of Practice).

A.3.4 *Test piece support*

The rod-supported design of Fig. A.8 is recommended (Section 6.4 of the Code of Practice).

A.3.5 *Test environment*

For COTEST, dry synthetic air containing 0.5 vol% SO_2 humidified to contain 2.5 vol% H_2O is recommended. If difficulties are encountered with the formation of H_2SO_4 in the furnace, the SO_2 or the H_2O may be omitted (Section 6.5 of the Code of Practice).

A.3.6 *Definition of deposit*

For COTEST, a salt mixture of $Na_2SO_4 : K_2SO_4$ corresponding to the eutectic composition with a theoretical melting point of 830 °C (75 mol% Na_2SO_4 and 25 mol% K_2SO_4) shall be applied at a target level of 0.6 mg/cm^2.

A.3.7 *Method of deposit application*

For COTEST, it is recommended that the deposit application be made by dissolving 1 wt% of the deposit according to 2.6 in water and spraying the salt solution obtained with a small paint sprayer using N_2 (or compressed air) as a carrier gas. The target amount of deposit is 0.6 mg/cm^2. Deposit should cover the whole specimen surface. Detailed instructions for deposit application on rectangular samples are given as follows.

Deposit application is made by spraying with a small paint sprayer using N_2 (or compressed air) as a carrier gas Fig. A.16. Deposit should cover the whole specimen surface. The test specimens are preheated at 150–200 °C for 10–15 min in their alumina support crucible to obtain an even deposit layer on the specimen surface. Only the specimens are taken out (with tweezers) from the preheating furnace to prevent unnecessary cooling. The preheating furnace should be located near (within 1 to 5 m) to the spraying set-up. The target amount of deposit is 0.6 mg/cm^2.

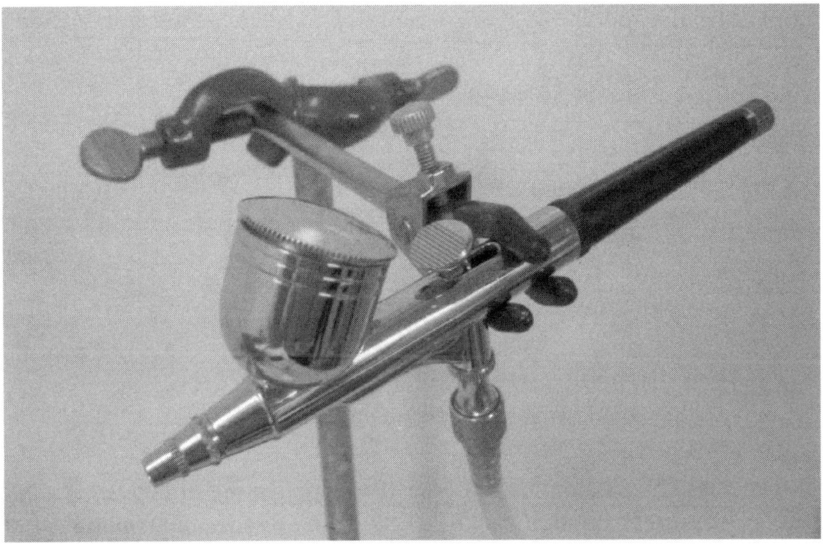

A.16 Small paint sprayer useful in deposit applications.

Before the deposit spraying, test sprayings with different spray distances (distance a in Fig A.17) and times are done to ensure the uniformity of spray and the adequate, even coverage area. The right distance can be determined by spraying the deposit on a piece of colourful paper at different distances *a*. If the deposit dries up quickly, it might block the paint sprayer. Therefore, the behaviour of the deposit should be observed during the distance tests. Test sprayings are also done to preheated samples to determine the spraying time. The time should not exceed the point where liquid deposit droplets start to form on the sample surface. If necessary, the sample can be reheated for a few minutes before spraying the other side of the sample.

Specimens are weighed at room temperature before and after the deposit application to obtain required amount of deposit. The spraying step and weighing are repeated if necessary.

Prepared specimens should be stored overnight in a desiccator to ensure a complete drying of deposit.

A.3.8 Frequency of reapplication

For COTEST, the deposit shall be reapplied every five cycles (once a week).

A.3.9 Determination of mass change

In addition to the intermediate mass change measurements described in the Code of Practice (Section 6.7.2), the mass of the test piece supporter with

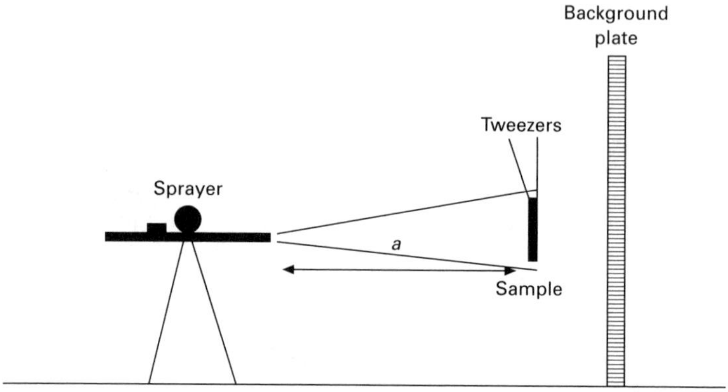

A.17 Schematic image of the spraying set-up.

test piece held by an alumina rod *after* the reapplication of deposit (m_{STD}), shall be measured. In addition, the mass of the test piece supporter after completion of the test, after any spall has been taken out from the test piece supporter by applying gentle mechanical means, such as brushing (m_{TDF}), shall be measured.

A.3.10 Frequency of mass change determination

Mass measurements at the completion of each cycle are recommended for the first five cycles. Thereafter the frequency may be reduced to every two or three cycles, except that measurements must be made before and after the reapplication of deposit (DEP) (Section 6.7.3 of the Code of Practice).

A.3.11 Analysis of mass change

The guidelines of the Code of Practice (Section 6.7.4 of the Code of Practice) hold regarding that values of mass changes used in the analysis shall be calculated using modified Eq. 6.1_{DEP}–6.4_{DEP} as follows:

6.7.2.4

Gross mass change Δm_{gross} as defined in 3.7 of the Code of Practice is determined according to Eq. 6.1_{DEP}.

$$\Delta m_{gross}(t_1) = m_{ST}(t_1) - m_{ST}(t_0) - m_{DEP}(t_1) \qquad 6.1_{DEP}$$

where $\Delta m_{gross}(t_1)$ = gross mass change at time t_1 (mg),
$m_{ST}(t_1)$ = mass of test piece supporter and test piece at time t_1 (mg),
$m_{ST}(t_0)$ = mass of test piece supporter and test piece prior to the deposit applications (mg),
$m_{DEP}(t_1)$ = calculated cumulative mass of deposit by time t_1 (mg); $m_{DEP}(t_1) = \Delta(m_{STD}(t_1) - m_{ST}(t_1))$.

6.7.2.5

The mass of spalled and that of any material eventually absorbed in the test piece supporter Δm_{spall} is determined according to Eq. 6.2_{DEP}:

$$\Delta m_{spall}(t_1) = m_S(t_1) - m_S(t_0) \qquad 6.2_{DEP}$$

where $\Delta m_{spall}(t_1)$ = mass change of spalled and absorbed material at time t_1 (mg),
$m_S(t_1)$ = mass of test piece supporter at time t_1 (mg),
$m_S(t_0)$ = mass of test piece supporter prior to the first deposit application (mg).

6.7.2.6 Net mass change Dm_{net} as defined in 3.8 is determined according to Eq. 6.3_{DEP} or 6.4_{DEP}.

$$\Delta m_{net}(t_1) = m_T(t_1) - m_T(t_0) - m_{DEP}(t_1) \qquad 6.3_{DEP}$$

where $\Delta m_{net}(t_1)$ = net mass change at time t_1 (mg),
$m_T(t_1)$ = mass of test piece at time t_1 (mg),
$m_T(t_0)$ = mass of test piece with no deposit applied prior to the test (mg),
$m_{DEP}(t_1)$ = calculated cumulative mass of deposit by time t_1 (mg); $m_{DEP}(t_1) = \Delta[m_{STD}(t_1) - m_{ST}(t_1)]$.

$$\Delta m_{net}(t_1) = \Delta m_{gross}(t_1) - \Delta m_{spall}(t_1) \qquad 6.4_{DEP}$$

where $\Delta m_{net}(t_1)$ = net mass change at time t_1 (mg),
$\Delta m_{gross}(t_1)$ = gross mass change at time t_1 (mg),
$\Delta m_{spall}(t_1)$ = mass of spalled material at time t_1 (mg).

A.4 Post-test evaluation of test pieces

The methodology described in section 7 of the Code of Practice shall apply. It is recommended that non-aqueous media be used for the grinding/polishing of cross-sections to avoid loss of water-soluble deposit/oxidation product.

A.5 Report

The information described in section 8 of the Code of Practice shall be reported. In addition, the following shall be reported:

8.1.2

- The weight of the deposit applied at each application (in mg/cm^2).
- The calculated total mass of deposit (mg); $m_{DEP} = \Delta[m_{STD}(t_1) - m_{ST}(t_1)]$, where the summation is taken over all the times t_1.
- The calculated mass change of the test piece supporter (mg); $\Delta m_{SD} = m_S(t_o) - m_{SDF}$.

8.1.3 The composition of the test gas.

10 Annex B: Test method for testing in low-velocity burner rigs

B.1 Scope

B.2 Test apparatus/principle of test

B.2.1

Low-velocity burner rig testing allows cyclic oxidation to be carried out with the additional influence of continuous deposition of salts. The test involves injection of fuel and salt solution into a heated chamber where combustion occurs. The combined action of the injection and the flow of the combustion atmosphere then carry the salt downstream where it may react with the fuel and/or vaporise and is then subsequently deposited/condensed onto the test pieces. The combustion and stabilisation zones of the burner rig are maintained at a higher temperature than the test piece zone in order to provide the driving force for deposition. The test pieces are rotated to ensure homogeneous deposition of salt around their circumference.

B.3 Test pieces

B.3.1

Test pieces shall, subject to limitations in available material, be of circular section and have a domed upper surface avoiding any sharp edges. Grooves/notches are allowable to secure the specimen in the holder. A suitable geometry is shown in Fig. A.18.

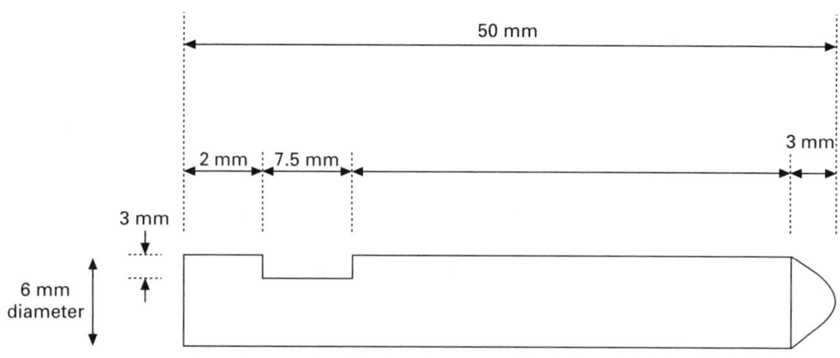

A.18 Burner rig specimen.

B.4 Test conditions

B.4.1

The test is defined by the following parameters:

- Temperature cycle
- Chemical composition of fuel and salt
- Fuel/air ratio
- Salt deposition rate

B.5 Thermocycles in burner rig testing

B.5.1

According to the definition in section 6.1.1 of the Code of Practice, a thermocycle consists of the heating phase, the hot dwell time, the cooling time and the cold dwell time. An example is given in Fig. A.19 for a hot dwell temperature of 1273 K. The four phases of a temperature cycle are defined in B.5.2 to B.5.5.

B.5.2

The heating time starts when the test pieces are heated, e.g. by entering the furnace, and ends with the beginning of the hot dwell time which is defined in B.5.3.

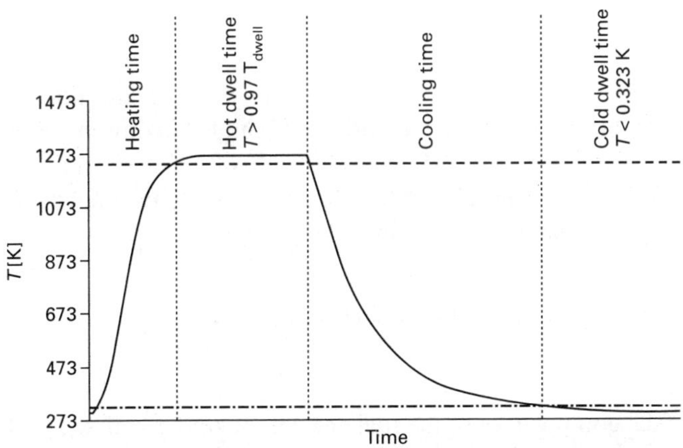

A.19 Definition of a thermocycle with a hot dwell time of 1273 K; dashed line with $T = 323\,K = 50\,°C$; dashed/dotted line with $T = 0.97\ T_{dwell} = 1234\,K$.

B.5.3

The hot dwell time starts when the actual temperature exceeds 97% of the desired hot dwell temperature T_{dwell}. Extensive numerical calculations and comparison between hypothetical and real temperature cycles have shown that only those times of the temperature cycle contribute to oxidation of the test pieces where the temperature is close to the hot dwell temperature (see Chapter 5). The hot dwell time ends upon removal from the furnace.

B.5.4

The cooling time starts when the heating of the test piece is stopped, e.g. by the removal of the test piece from the furnace, and ends when the actual test piece temperature falls below 50 °C.

B.5.5

The cold dwell time starts after the test pieces have cooled below 50 °C and ends when the test pieces are heated again.

B.5.6

For burner rig testing the cycle is constrained by a daily/weekly cycle.

B.5.7

The recommended cycle is 20 h hot dwell and 4 h total for cooling time, cold dwell and heating time.

B.5.8

The first cycle of the week should start on Monday afternoon, allowing specimens to be removed for measurements the following morning. Unless weekend working is envisaged, the final cycle of the week should finish on Friday morning. After measurement the specimens shall be stored in a desiccator until cycling recommences the following Monday.

B.6 Fuel supply and salt deposition

B.6.1

The chemical composition of the fuel and salt, together with the fuel/air ratio and salt deposition rate, shall be agreed with the customer prior to commencement of the test programme.

B.6.2

Fuel shall be delivered continuously at the required rate for at least 15 mins prior to introduction of the specimens to the chamber and throughout the exposure. The fuel may be switched off after removal of the specimens from the chamber

B.6.3

The deposition rate of the salt is controlled by the contaminant flux rate, salt composition and temperature. The contaminant flux rate for the required deposition rate test conditions shall be established prior to the experiment using inert pins in the specimen chamber.

B.6.4

Deposition during the exposure shall be monitored using an inert pin positioned in the specimen holder together with the testpieces.

B.7 Testing duration

The testing duration shall be at least 300 h of accumulated hot dwell time to allow a significant oxidation of the test pieces. For more reliable results it is, however, recommended to extent the accumulated hot dwell time to 1000 h.

B.8 Supporting of test pieces

The test pieces shall be supported according to the following principles:

B.8.1

The test piece shall be supported by a material that does not react at the test temperature.

B.8.2

The specimens shall be continuously rotated to ensure even distribution of deposit around the circumference of the specimens.

B.8.3

Additional restraint in the form of heat resistant wire may be necessary to secure the test pieces in the rotating holder.

B.9 Determination of mass change by oxidation

B.9.1

Test pieces shall never be touched with the hands in order to exclude any contamination (fat, salts). Care has to be taken when using gloves as the contamination with the separating agent of the gloves leads to falsification in mass determination. The use of tweezers is recommended.

B.9.2

The mass of the test pieces shall be measured at predetermined intervals. Measurement shall be more frequent at the early stages of the test.

B.9.3

The mass change due to oxidation/corrosion is calculated by subtracting the mass change due to salt deposition from the total mass change. The mass of salt deposited is measured on the inert pins that are exposed in parallel with the test pieces.

B.9.4

Some test pieces may be washed after mass measurements. In this case the mass change prior to and after washing shall be recorded.

B.10 Metallographic examination post-test evaluation of test pieces

The methodology described in section 7 of the Code of Practice shall apply. It is recommended that non-aqueous media be used for the grinding/polishing of cross-sections to avoid loss of water-soluble deposit/oxidation product.

B.11 Reporting of results

The information described in section 8 of the Code of Practice shall be reported. In addition, the following shall be reported:

B.11.1 Testing environments

- Composition of fuel.

Index